WJEC
Mathematics
for AS Level

Pure & Applied Revision Guide

Stephen Doyle

Illuminate
Publishing

Published in 2020 by Illuminate Publishing Limited, an imprint of Hodder Education, an Hachette UK Company, Carmelite House, 50 Victoria Embankment, London EC4Y 0DZ

Orders: please contact Hachette UK Distribution, Hely Hutchinson Centre, Milton Road, Didcot, Oxfordshire, OX11 7HH. Telephone: +44 (0)1235 827827. Email: education@hachette.co.uk. Lines are open from 9 a.m. to 5 p.m., Monday to Friday. You can also order through our website: www.hoddereducation.co.uk

© Stephen Doyle 2020

The moral rights of the author have been asserted.

All questions in this book are © WJEC and taken from past exam papers.

All rights reserved. No part of this book may be reprinted, reproduced or utilised in any form or by any electronic, mechanical, or other means, now known or hereafter invented, including photocopying and recording, or in any information storage and retrieval system, without permission in writing from the publishers.

British Library Cataloguing in Publication Data

A catalogue record for this book is available from the British Library

ISBN 978-1-912820-33-7

Printed by Ashford Colour Press, UK

Impression 4
Year 2024

Hachette UK's policy is to use papers that are natural, renewable and recyclable products and made from wood grown in well-managed forests and other controlled sources. The logging and manufacturing processes are expected to conform to the environment regulations of the country of origin.

Editor: Geoff Tuttle
Cover design: Neil Sutton
Text design and layout: GreenGate Publishing Services, Tonbridge, Kent

Photo credits

Cover: Klavdiya Krinichnaya/Shutterstock; **p11** Radachynskyi Serhii/Shutterstock; **p14** Anatoli Styf/Shutterstock; **p30** Jeeraphun Kulpetjira/Shutterstock; **p41** Jan Miko/Shutterstock; **p45** Africa Studio/Shutterstock; **p53** Fotos593/Shutterstock; **p62** jo Crebbin/Shutterstock; **p71** Jose Antonio Perez/Shutterstock; **p78** Rena Schild/Shutterstock. **p84** Marco Ossino/Shutterstock; **p90** Sergey Nivens/Shutterstock; **p106** Blackregis/Shutterstock; **p113** urickung/Shutterstock; **p122** Gajus/Shutterstock; **p132** Will Rodrigues/Shutterstock; **p141** Andrea Danti/Shutterstock; **p147** sirtravelalot/Shutterstock.

Acknowledgements

The author and publisher wish to thank Sam Hartburn for her careful attention when reading this book.

FSC™
www.fsc.org

MIX
Paper | Supporting responsible forestry
FSC™ C104740

Contents

How to use this book   4

AS Unit 1 Pure Mathematics A

Topic 1 Proof   11
Topic 2 Algebra and functions   14
Topic 3 Coordinate geometry in the (x, y) plane   30
Topic 4 Sequences and series – the binomial theorem 41
Topic 5 Trigonometry   45
Topic 6 Exponentials and logarithms   53
Topic 7 Differentiation   62
Topic 8 Integration   71
Topic 9 Vectors   78

AS Unit 2: Applied Mathematics A

Section A: Statistics
Topic 10 Statistical sampling   84
Topic 11 Data presentation and interpretation   90
Topic 12 Probability   106
Topic 13 Statistical distributions   113
Topic 14 Statistical hypothesis testing   122

Section B: Mechanics
Topic 15 Kinematics   132
Topic 16 Dynamics of a particle   141
Topic 17 Vectors   147

Answers   151

How to use this book

The purpose of this book is to aid revision just before the examination, so it is assumed you have completed the course and have a reasonable knowledge of the topics. The book has been specifically written for the WJEC AS Level course and has been produced by an experienced author and teacher. The book includes full coverage of both the Pure and Applied Mathematics units, so it covers the entire AS course.

A feature of the new exams is the lack of scaffolding in some questions. When questions have scaffolding, they guide you through the steps by questioning you on each step as you go along until you arrive at a final answer. The same question without scaffolding will not ask you intermediate questions but will instead just ask you to find the final answer. Part of the process of answering them involves understanding the question and planning the steps involved in finding the solution. However, not all questions are like this. Some are more traditional in that there is scaffolding, and it is fairly easy to see what needs to be done to answer them. In this book we will cover both types of questions.

Other maths revision books don't explore this at all and treat the answering of these questions in the same way as you would with the old specification.

Like most revision books, we will go through worked examples, but you will be encouraged to analyse the question before you start and be able to plan what you have to do.

As part of the worked examples, there will be tips which will enable you to see the way forward with answers. Also included will be 'watch outs' which will be points or misconceptions made by students.

Knowledge and understanding

Topics start with a list of material you need to know from your GCSE studies. Students frequently trip up on material such as algebra and fractions, so if you have any weaknesses in the GCSE material, you need to look at the material again.

There is then a section called 'Quick revision' which summarises the material you need to know.

Looking at exam questions

Answering full examination questions lies at the heart of this book so rather than look at small parts of questions, we will instead be concentrating on complete examination questions.

Rather than go straight to the answers, we will instead look at the analysis of the question and the thought processes involved in deciding how to answer the question.

Each question is divided into the following parts:

- The question
- Thinking about the question – what does the question involve? What specification topics does it involve?
- Starting the solution – planning the steps you need to take to answer the questions
- The solution.

The examination questions are also annotated with advice about the knowledge, skills and techniques needed to answer them.

Exam practice

These questions are of examination standard and they allow you to have a go at them on your own. Full solutions are provided and they show you the steps to take.

Planning your revision and effective ways to revise

Revision is very much a personal thing – some people revise more effectively listening to music, while others need complete silence. Here are a few tips that may work for you:

Check you have completed all the material in each topic – check you have completed all the topics in the specification by printing out a copy of the specification from the WJEC website and match the material against your notes.

Ensure sure you understand the basics – fractional indices, the solution of simultaneous and quadratic equations, completing the square, etc. Remember that your knowledge of higher-level GCSE Maths will be assumed.

Target your revision – have a firm understanding of what you know and evaluate your strengths and weaknesses. Don't spend too much time going through things you know, but instead concentrate on your weaknesses.

Maximise your time – try to concentrate on your revision without breaking off to check your email, social media, etc. You may find it easier to work in a library where you won't be distracted.

Use technology – if there is no teacher to ask about a solution to a question or to ask about a particular topic, use sites such as YouTube where there are videos on many different AS Maths topics.

Use your time effectively – use your phone to check formulae you need to remember or for explanations of topics you are unsure of during any spare time you have.

The use of calculators

All the papers in AS Maths allow you to use a calculator. The calculator you used for your GCSE or IGCSE maths will probably not have the statistical functions you need for the Applied part of the course. You can avoid purchasing a more advanced calculator containing statistical functions and instead use statistical tables provided free; however, the ease in using calculators massively outweighs the cost of them. Using the advanced calculator, you will be able to compute summary statistics and access probabilities from standard statistical distributions.

Which calculator is recommended – you have to be careful that the one you buy is allowed by the examination board. Your teacher/lecturer probably recommended one for you and the most popular one by far is the CASIO fx-991EX ClassWiz.

A bit of advice, though, don't purchase a new calculator just before the exam. You need practice at using it. Also, it might be worth buying a book which explains how to use the calculator in a way that is useful for A-Level Maths or look at some of the many videos on YouTube. You will find lots of useful shortcuts and ways of using the calculator that will enable you to save time as well as improve your accuracy when answering questions.

Information about your examination

There are two units and both must be taken.

AS Unit 1: Pure Mathematics A – a written examination 2 hours 30 minutes and worth 120 marks.

AS Unit 2: Applied Mathematics A – a written examination 1 hour 45 minutes and worth 75 marks. This unit is divided into two sections:

> It is up to you how you divide your time between sections A and B.

　　Section A: Statistics (40 marks)

　　Section B: Mechanics (35 marks)

Calculators can be used in the exam for both the units.

Formulae and identities you must remember

Some formulae and identities are included on the formula sheet. Here are the formulae and identities that are not included on the formulae sheet and therefore must be remembered.

Quadratic equations

$ax^2 + bx + c = 0$ has roots $\dfrac{-b \pm \sqrt{b^2 - 4ac}}{2a}$

Laws of indices

$$a^x \times a^y = a^{x+y}$$

$$a^x \div a^y = a^{x-y}$$

$$(a^x)^y = a^{xy}$$

> Why not take a picture of all these formulae and equations on your phone and keep referring to them. Hopefully, after a while you will remember them.

Laws of logarithms

$$x = a^n \iff n = \log_a x \text{ for } a > 0 \text{ and } x > 0.$$

$$\log_a x + \log_a y \equiv \log_a xy$$

$$\log_a x - \log_a y \equiv \log_a \left(\dfrac{x}{y}\right)$$

$$k \log_a x \equiv \log_a x^k$$

Coordinate geometry

A straight-line graph, gradient m passing through (x_1, y_1) has equation

$$y - y_1 = m(x - x_1)$$

Straight lines with gradients m_1 and m_2 are perpendicular when

$$m_1 m_2 = -1$$

Trigonometry

In the triangle ABC

Sine rule: $\quad \dfrac{a}{\sin A} = \dfrac{b}{\sin B} = \dfrac{c}{\sin C}$

Cosine rule: $\quad a^2 = b^2 + c^2 - 2bc \cos A$

$$\text{Area} = \tfrac{1}{2} ab \sin C$$

$$\cos^2 \theta + \sin^2 \theta = 1$$

Mensuration

Circumference and area of circle, radius r and diameter d:

$$C = 2\pi r = \pi d \qquad A = \pi r^2$$

Pythagoras' theorem: In any right-angled triangle where a, b and c are the lengths of the sides and c is the hypotenuse:

$$c^2 = a^2 + b^2$$

Area of trapezium = $\frac{1}{2}(a + b)h$, where a and b are the lengths of the parallel sides and h is their perpendicular separation.

Volume of a prism = area of cross section × length

For a circle of radius, r, where an angle at the centre of θ radians subtends an arc of length s and encloses an associated sector of area A:

$$s = r\theta \qquad A = \frac{1}{2}r^2\theta$$

Differentiation

Function	*Derivative*
x^n	nx^{n-1}

Integration

Function	*Integral*
x^n	$\frac{1}{n+1}x^{n+1} + c, n \neq 0$

$$\text{Area under a curve} = \int_a^b y \, dx \quad (y \geq 0)$$

Mechanics

Forces and equilibrium

$$\text{Weight} = \text{mass} \times g$$

Newton's second law in the form: $F = ma$

Kinematics

For motion in a straight line with variable acceleration:

$$v = \frac{dv}{dr} \qquad\qquad a = \frac{dv}{dt} = \frac{d^2r}{dt^2}$$

$$r = \int v \, dt \qquad\qquad v = \int a \, dt$$

Statistics

$$\bar{x} = \frac{\sum x}{n} = \frac{\sum fx}{\sum f}$$

Formulae included on the formula sheet

The following formulae need not be remembered as they are included on the formula booklet that will be given to you in the exam.

The formulae here are all the ones applicable to AS. In the actual formula booklet they will be mixed up with other formulae needed for A2 or further maths, so make sure you familiarise yourself with the actual booklet.

Mensuration

Surface area of sphere = $4\pi r^2$

Area of curved surface of cone = $\pi r \times$ slant height

Binomial series

$$(a + b)^n = a^n + \binom{n}{1}a^{n-1}b + \binom{n}{2}a^{n-2}b^2 + \ldots + \binom{n}{r}a^{n-r}b^r + \ldots + b^n \quad (n \in \mathbb{N})$$

where $\binom{n}{r} = {}^nC_r = \dfrac{n!}{r!(n-r)!}$

$$(1 + x)^n = 1 + nx + \frac{n(n-1)}{1 \times 2}x^2 + \ldots + \frac{n(n-1)\ldots(n-r+1)}{1 \times 2 \times \ldots \times r}x^r \quad (|x| < 1, n \in \mathbb{N})$$

Vectors

The point dividing AB in the ratio $\lambda : \mu$ is $\dfrac{\mu \mathbf{a} + \lambda \mathbf{b}}{\lambda + \mu}$

Probability and statistics

Probability

$$P(A \cup B) = P(A) + P(B) - P(A \cap B)$$

Standard discrete distributions

Distribution of X	$P(X = x)$	Mean	Variance
Binomial B(n, p)	$\binom{n}{x}p^x(1-p)^{n-x}$	np	$np(1-p)$
Poisson Po(λ)	$e^{-\lambda}\dfrac{\lambda^x}{x!}$	λ	λ

Sampling distributions

For a random sample $X_1, X_2, \ldots, X_n$ of n independent observations from a distribution having mean μ and variance σ^2

$\overline{X}$ is an unbiased estimator of μ, with Var$\left(\overline{X}\right) = \dfrac{\sigma^2}{n}$

S^2 is an unbiased estimator of σ^2, where $S^2 = \dfrac{\sum(X_i - \overline{X})^2}{n-1}$

Tips for the actual exam

Do not change your calculator to a different model just before your exam. Make sure you understand how to use all the functions, especially the statistical functions, as the exam itself is not the place to start learning them.

Pace yourself and avoid getting bogged down on a question. Note that you do not have to do questions in the order they are presented on the exam paper.

Include answer steps with clear explanations.

If you feel confident – include shortcuts to workings out – however, remember that time saved sometimes comes at the expense of accuracy.

Tidy up final answers – remember to cancel fractions, simplify algebraic expressions, give answers to an appropriate accuracy, etc.

Draw clearly labelled diagrams, if appropriate. Always draw graphs for coordinate geometry questions even if they are not asked for.

Watch out when drawing curves with asymptotes that the curves do not touch the asymptotes.

In trigonometric proofs, do not work with both sides together. It is best to start with the more complicated side on its own and then prove that it is equal or equivalent to the less complicated side.

For example to prove

$$\frac{\sin^3 \theta + \sin \theta \cos^2 \theta}{\cos \theta} \equiv \tan \theta$$

you would start with the more complicated side (i.e. the left-hand side) and manipulate it to prove the less complicated right-hand side.

Make it clear what your answer is – the examiner should not have to search for your answer amongst a load of working out.

Good luck with your revision and for the exam itself.

Steve Doyle

1 Proof

Prior knowledge

You will need to make sure you fully understand the following from your GCSE studies:

- Understanding and using equations and formulae

Quick revision

Types of numbers

Real numbers – numbers that give a positive number when squared.

Imaginary numbers – numbers that, when squared, give a negative number (e.g. $\sqrt{-1}$ when squared gives -1).

Rational numbers – numbers that can be expressed as a fraction.

Watch out

Proofs are not popular with students, as they can be so different. Sometimes it is hard to know where to start so it is one of the topics you will really need to work at. You need to do as many questions as you can. The main thing is to persevere with the topic.

Types of proof

Proof by exhaustion

Use all the allowable values to prove that a mathematical statement is true or false. Only use this proof if there are only a small number of possible values to try.

Disproof by counter-example

This proof is used to prove a property is false by providing an example where it does not hold. If the example does not hold for one case then it cannot be true for all cases.

Proof by deduction

This proof uses something already known or assumed to be true. In many cases it uses algebra to decide whether a statement is true or false.

Looking at exam questions

1 In each of the two statements below, c and d are real numbers. One of the statements is true, while the other is false.

A: $(2c - d)^2 = 4c^2 - d^2$, for all values of c and d.

B: $8c^3 - d^3 = (2c - d)(4c^2 + 2cd + d^2)$, for all values of c and d.

(a) Identify the statement which is false. Show, by counter-example, that this statement is in fact false. [2]

(b) Identify the statement which is true. Give a proof to show that this statement is in fact true. [2]

Thinking about the question

Proving a statement is true cannot be done by simply substituting in numbers for the variables and seeing if the left-hand side of the statement is equal to the right-hand side. You would only be proving it for the values you used and not for all possible values.

We therefore need to find values for a counter-example for the statement that is false.

Starting the solution

We could start by using algebra to prove the statement that is true using proof by deduction.

You can start by multiplying out the brackets on the left-hand side of statement A and see if it equals the expression on the right-hand side. You can then multiply out the brackets on the right-hand side of B and see if it equals the expression on the left-hand side.

It is important to note that you cannot use proof by deduction for the statement that is false as the question asks that disproof by counter-example is used. So, for part (a) we need to find values where the right-hand and left-hand sides are no longer equal.

You need to experiment by substituting different numbers for c and d until you find a pair of values where the expressions are not equal.

The solution

(a) Let $c = 2$ and $d = 1$.

$(2c - d)^2 = (4 - 1)^2 = 9$ and $4c^2 - d^2 = 16 - 1 = 15$

Since $9 \neq 15$ we have a counter-example so statement A is false.

(b) $(2c - d)(4c^2 + 2cd + d^2) = 8c^3 + 4c^2d + 2cd^2 - 4c^2d - 2cd^2 - d^3$

$$= 8c^3 - d^3$$

Hence statement B is true.

This is a proof by deduction as you are using algebra to decide whether the statement is true or false.

Exam practice

1. (a) Given that n is an even number, prove that the expression $10n^2 + 5n$ has a factor of 10. [2]
   (b) Use disproof by counter-example to determine if $10n^2 + 5n$ has a factor of 10 for any integer n. [2]

2. Using disproof by counter-example, prove that the proposition '$\sqrt{xy} \leq \frac{1}{2}(x + y)$' is not true for all real values of x and y. [4]

3. Prove that the sum of three consecutive odd numbers is also an odd number. [3]

4. Use proof by exhaustion to prove the following statement: 'Every perfect cube number is a multiple of 9, one less than a multiple of 9 or one more than a multiple of 9.' [5]

5. Show, by counter-example that the statement 'If $\cos \theta = \cos \varphi$ then $\sin \theta = \sin \varphi$' is false. [4]

6. Show, by counter-example, that the following statement is false. 'If the integers a, b, c, d are such that a is a factor of c and b is a factor of d, then $(a + b)$ is a factor of $(c + d)$.' [3]

7. Prove that for all real values of n, $n^2 - 2n + 2$ is positive. [3]

8. Using a suitable proof, prove that if n is a positive integer, then $n^2 + 2$ is *not* a multiple of 4 for $2 \leq n \leq 4$. [5]

2 Algebra and functions

Prior knowledge

You will need to make sure you fully understand the following from your GCSE studies:

- The laws of indices

- Rationalisation of surds

- Solving quadratic equations by factorisation or using the formula

- Solving simultaneous equations

- Drawing straight lines and curves from their equations

- Interpreting and applying the transformation of functions graphically including:
  $y = f(x + a)$, $y = f(kx)$, $y = kf(x)$ and $y = f(x) + a$, applied to $y = f(x)$.

Quick revision

Laws of indices

$$a^m \times a^n = a^{m+n}$$

$$a^m \div a^n = a^{m-n}$$

$$(a^m)^n = a^{m \times n}$$

$$a^{-m} = \frac{1}{a^m}$$

If $a \neq 0$, $a^0 = 1$

$$a^{\frac{m}{n}} = \sqrt[n]{a^m} = \left(\sqrt[n]{a}\right)^m$$

$$a^{-\frac{m}{n}} = \frac{1}{a^{\frac{m}{n}}} = \frac{1}{\sqrt[n]{a^m}} \text{ or } \frac{1}{\left(\sqrt[n]{a}\right)^m}$$

Examples

$$2^2 \times 2^2 \times 2^2 = 2^6$$
$$3^5 \div 3^2 = 3^3$$
$$(5^4)^2 = 5^8$$

Examples

$$2^{-1} = \frac{1}{2^1} = \frac{1}{2}$$
$$8^{\frac{2}{3}} = \sqrt[3]{8^2} = 4$$
$$27^{-\frac{2}{3}} = \frac{1}{27^{\frac{2}{3}}} = \frac{1}{\sqrt[3]{27^2}} = \frac{1}{9}$$

Surds

$$\sqrt{a} \times \sqrt{a} = a$$

$$\sqrt{a} \times \sqrt{b} = \sqrt{ab}$$

$$\left(\sqrt{a} + \sqrt{b}\right)\left(\sqrt{a} - \sqrt{b}\right) = a - b$$

Examples

$$\sqrt{5} \times \sqrt{5} = 5$$
$$\sqrt{6} \times \sqrt{5} = \sqrt{30}$$
$$\left(\sqrt{3} + \sqrt{2}\right)\left(\sqrt{3} - \sqrt{2}\right)$$
$$= 3 - 2 = 1$$

Simplifying surds

A surd is in its simplest form when there are no square factors in the number inside the square root $\left(\text{e.g. } \sqrt{50} = \sqrt{25 \times 2} = 5\sqrt{2}\right)$.

Rationalisation of surds

This involves avoiding having surds in the denominator. We avoid having surds in the denominator and removing them is called rationalising the denominator.

$$\frac{a}{b\sqrt{c}} = \frac{a}{b\sqrt{c}} \times \frac{\sqrt{c}}{\sqrt{c}} = \frac{a\sqrt{c}}{bc}$$

(Here the denominator is rationalised by multiplying the top and bottom by $\sqrt{c}$.)

$$\frac{a}{\sqrt{b} \pm \sqrt{c}} = \frac{a}{\sqrt{b} \pm \sqrt{c}} \times \frac{\sqrt{b} \mp \sqrt{c}}{\sqrt{b} \mp \sqrt{c}} = \frac{a\sqrt{b} \mp a\sqrt{c}}{b - c}$$

(Here the denominator is rationalised by multiplying the top and bottom of the expression by the conjugate of the denominator.)

Completing the square

Completing the square involves writing a quadratic expression such as $x^2 + 6x + 8$ in the form $(x + p)^2 + q$.

To complete the square for $x^2 + 6x + 8$

1   Halve the coefficient of x, add it to x and then bracket and square, i.e. $(x + 3)^2$.

2   Square the number you have added (i.e. 3^2) and subtract from the expression and then add the number originally present, i.e. 8 in this case.

Hence $x^2 + 6x + 8 = (x + 3)^2 - 9 + 8 = (x + 3)^2 - 1$

To complete the square for $3x^2 + 5x - 9$ you must take the coefficient of x^2 out as a factor like this:

$$3x^2 + 5x - 9 = 3\left[x^2 + \frac{5}{3}x - 3\right].$$

Then complete the square for the expression in the square bracket and then multiply by 3.

Hence $3\left[x^2 + \frac{5}{3}x - 3\right] = 3\left[\left(x + \frac{5}{6}\right)^2 - \frac{25}{36} - 3\right]$

$$= 3\left[\left(x + \frac{5}{6}\right)^2 - \frac{133}{36}\right] = 3\left(x + \frac{5}{6}\right)^2 - \frac{133}{12}\right]$$

Solving quadratic equations

Equations in the form $ax^2 + bx + c$ can be solved:

1   By factorising. You should be familiar with this from your GCSE work.

2   By completing the square and equating the result to zero.

3   By using the quadratic formula $x = \dfrac{-b \pm \sqrt{b^2 - 4ac}}{2a}$.

The discriminant of a quadratic function

The roots of the quadratic equation $ax^2 + bx + c$ are the solutions and also the x-coordinates of the points where the curve of the function cuts the x-axis.

$b^2 - 4ac$ is the discriminant

If $b^2 - 4ac > 0$, then there are two real and distinct (i.e. different) roots.

If $b^2 - 4ac = 0$, then there are two real and equal roots.

If $b^2 - 4ac < 0$, then there are no real roots.

Sketching the graph of a quadratic function to find the maximum or minimum point

Complete the square so the equation is in the form, $y = a(x + p)^2 + q$.

If a is positive (i.e. $a > 0$) the curve will be ∪-shaped and if a is negative (i.e. $a < 0$) the curve will be ∩-shaped.

The vertex (i.e. the maximum or minimum point) will be at $(-p, q)$.

The axis of symmetry will be $x = -p$

Representing linear and quadratic inequalities graphically

Strict inequality – where the expression is < or > a value.

Non-strict inequality – where the expression is ≤ or ≥ a value.

To represent linear and quadratic inequalities graphically:

Use the equations to enable you to draw the graphs. Draw the graphs of the equations with equals signs, as it is the area that represents the inequality.

For example, if we had to shade the area represented by $y \geq x^2 + 2$, $x > -1$ and $x + y \leq 4$ we would draw the following graph.
As $y \geq x^2 + 2$ the region would be above this curve, for $x \geq -1$ the region would be to the right of this line and for $x + y \leq 4$ the region would be below this line. Hence the region satisfying all three inequalities can now be shaded as shown.

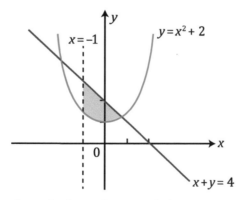

Set notation for the solution of inequalities

Instead of writing a solution such as $0 < x \leq 3$, if you were asked to write the solution using set notation, it would be written as $\{x : 0 < x \leq 3\}$.

Solving quadratic inequalities

1   Equate the quadratic to zero and then solve to find the two critical values.

2   Do a quick sketch of the curve.

3   Decide whether the part (or parts) of the curve is above or below the x-axis (i.e. if $\leq$ the required region will be on and below the x-axis and if $\geq$ it will be on and above the x-axis).

The remainder theorem: If a polynomial $f(x)$ is divided by $(x - a)$ the remainder is $f(a)$.

The factor theorem: For a polynomial $f(x)$, if $f(a) = 0$ then $(x - a)$ is a factor of $f(x)$.

Sketching curves of functions

This section looks at sketching curves defined by simple functions. You must be able to spot these simple functions and be able to immediately recognise the shape of the graph.

Graphs of $y = x^2$, $y = x^4$ and other even powers of x

These graphs are ∪-shaped and pass through the origin and look like this.

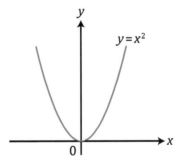

Graphs of $y = x^3$, $y = x^5$ and other odd powers of x

These graphs pass through the origin and look like this.

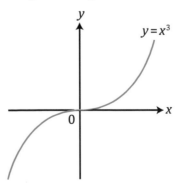

Graph of $y = \dfrac{a}{x}$ (i.e. reciprocal graphs)

The graphs of $y = \dfrac{a}{x}$, where a is a constant all take a similar shape regardless of the value of a. These graphs all have the x and y axes as asymptotes.

The straight line which a graph approaches without actually touching is called an asymptote.

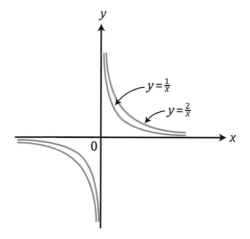

Notice as the value of a increases, the graphs have a similar shape but they lie further from the origin but still have both axes as asymptotes.

Graph of $y = \dfrac{a}{x^2}$ (i.e. reciprocal graphs)

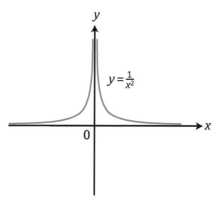

All the graphs of $y = \dfrac{a}{x^2}$ have the x-axis and the positive y-axis as asymptotes. As the value of a increases then so does the distance of the graph from the origin.

Transformations of curves

Transformations of the graph of $y = f(x)$

Original function	New function	Transformation
$y = f(x)$	$y = f(x) + a$	Translation of a units parallel to the y-axis, i.e. translation of $\begin{pmatrix} 0 \\ a \end{pmatrix}$
$y = f(x)$	$y = f(x + a)$	Translation of a units to the left, parallel to the x-axis, i.e. translation of $\begin{pmatrix} -a \\ 0 \end{pmatrix}$
$y = f(x)$	$y = f(x - a)$	Translation of a units to the right, parallel to the x-axis, i.e. translation of $\begin{pmatrix} a \\ 0 \end{pmatrix}$
$y = f(x)$	$y = -f(x)$	A reflection in the x-axis
$y = f(x)$	$y = af(x)$	One-way stretch with scale factor a parallel to the y-axis
$y = f(x)$	$y = f(ax)$	One-way stretch with scale factor $\frac{1}{a}$ parallel to the x-axis

$y = f(x)$ $\qquad$ $y = f(x) + a$

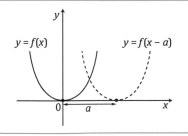

$y = f(x)$ $\qquad$ $y = f(x + a)$

$y = f(x)$ $\qquad$ $y = f(x - a)$

Original function	New function	Transformation
$y = f(x)$	$y = -f(x)$	
$y = f(x)$	$y = af(x)$ E.g. $y = 2f(x)$	
$y = f(x)$	$y = f(ax)$ E.g. $y = f(2x)$	

Formulae included and not included

Formulae included	Formulae not included
There are no formulae included in the formula booklet for this topic.	For $ax^2 + bx + c = 0$, $x = \dfrac{-b \pm \sqrt{b^2 - 4ac}}{2a}$
	For $ax^2 + bx + c = 0$, discriminant $= b^2 - 4ac$

Looking at exam questions

1 Showing all your working, simplify:

(a) $\dfrac{24\sqrt{a}}{(\sqrt{a}+3)^2-(\sqrt{a}-3)^2}$ [3]

(b) $\dfrac{3\sqrt{7}+5\sqrt{3}}{\sqrt{7}+\sqrt{3}}$ [4]

Thinking about the question

For part (a), we need to multiply out the brackets in the denominator and tidy up the result.

For part (b), we need to rationalise the denominator.

Starting the solution

1   After multiplying the brackets in the denominator and simplifying, we can then cancel the remaining algebraic fraction.

2   The conjugate of the denominator is $\sqrt{7}-\sqrt{3}$ so the numerator and denominator need to be multiplied by this.

The solution

> **Watch out**
> It would be easy to get a sign wrong when multiplying out these brackets.

(a) $\dfrac{24\sqrt{a}}{(\sqrt{a}+3)^2-(\sqrt{a}-3)^2}=\dfrac{24\sqrt{a}}{(a+6\sqrt{a}+9)-(a-6\sqrt{a}+9)}=\dfrac{24\sqrt{a}}{12\sqrt{a}}=2$

> Multiply the numerator and denominator by the conjugate of the denominator.

(b) $\dfrac{3\sqrt{7}+5\sqrt{3}}{\sqrt{7}+\sqrt{3}}=\dfrac{(3\sqrt{7}+5\sqrt{3})(\sqrt{7}-\sqrt{3})}{(\sqrt{7}+\sqrt{3})(\sqrt{7}-\sqrt{3})}$

$=\dfrac{21-3\sqrt{21}+5\sqrt{21}-15}{7-3}=\dfrac{6+2\sqrt{21}}{4}=\dfrac{1}{2}\left(3+\sqrt{21}\right)$

---

2 Express $4x^2-8x+6$ in the form $a(x+b)^2+c$, where the values of the constants a, b and c are to be found. [3]

Hence, sketch the graph of $y=4x^2-8x+6$, indicating the coordinates of its stationary point. [3]

Thinking about the question

$4x^2-8x+6$ is a quadratic function and the form given means we have to complete the square.

Completing the square enables the coordinates of the stationary point to be found.

As the coefficient of x^2 is positive, the curve will be ∪-shaped.

Starting the solution

As the coefficient of x^2 is 4, this number needs to be taken out as a factor. We then complete the square of the remainder and finally multiply the 4 back in.

As the curve is ∪-shaped, there will be a minimum point at $(-b, c)$. We can then draw the graph with this marked. Note the question has not asked for the coordinates of the points of intersection of the curve with the axes.

The solution

$$4x^2 - 8x + 6 = 4\left[x^2 - 2x + \tfrac{3}{2}\right]$$

$$= 4\left[(x-1)^2 - 1 + \tfrac{3}{2}\right]$$

$$= 4\left[(x-1)^2 + \tfrac{1}{2}\right]$$

$$= 4(x-1)^2 + 2$$

Hence, $a = 4$, $b = -1$ and $c = 2$

The stationary point (a minimum point in this case) will be at $(-b, c)$ which gives the point $(1, 2)$.

The graph is

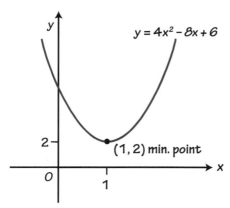

$y = 4x^2 - 8x + 6$

$(1, 2)$ min. point

> 4 is taken out as a factor before completing the square.

> Compare your answer with the format for the expression in the question, which in this case is $a(x + b)^2 + c$ to find the values of a, b and c.

> **Watch out**
> Make sure you label the origin, both axes, the equation of the curve and mark the minimum point and give its coordinates.

---

3   Find the values of m for which the equation
$4x^2 + 8x - 8 = m(4x - 3)$ has real roots.   [5]

Thinking about the question

1   This is a question about roots. For real roots of a quadratic equation, the discriminant, $b^2 - 4ac \geq 0$.

2   The equation given needs to be in the form of a quadratic equation that is equated to zero. We then arrange the equation in descending powers of x.

Starting the solution

1   Multiply out the bracket on the right of the equals.

2   Rearrange the equation so that it is in descending powers of x and is equal to zero.

3   Find an expression for the discriminant and put it equal to zero.

4   Solve the resulting equation in m.

The solution

$$4x^2 + 8x - 8 = m(4x - 3)$$

$$4x^2 + 8x - 8 = 4mx - 3m$$

$$4x^2 + 8x - 4mx + 3m - 8 = 0$$

$$4x^2 + (8 - 4m)x + (3m - 8) = 0$$

> Note that we arrange the equation in descending powers of x (i.e. x^2 first, then x and then just numbers).

For real roots of a quadratic equation, the discriminant,

$$b^2 - 4ac \geq 0.$$

> When you get large coefficients like this, there is usually a number that will go into them all. In this case we can divide both sides of the equation by 16.

Hence $(8 - 4m)^2 - 4\,(4)(3m - 8) \geq 0$

$$64 - 64m + 16m^2 - 48m + 128 \geq 0$$

$$16m^2 - 112m + 192 \geq 0$$

$$m^2 - 7m + 12 \geq 0$$

$$(m - 3)(m - 4) \geq 0$$

> It's always worth spending time drawing a quick sketch of the curve including the intercepts on the x-axis.

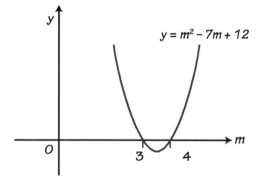

We want the sections of the curve on or above the m-axis.

Hence $m \leq 3$ or $m \geq 4$

---

4   Given that $(x - 2)$ and $(x + 2)$ are factors of the polynomial $2x^3 + px^2 + qx - 12$,

   (a)   find the values of p and q,   [4]

   (b)   determine the other factor of the polynomial.   [1]

Thinking about the question

For part (a) there are two unknowns, p and q and when this happens, we usually have to find two simultaneous equations to solve.

One way to find the other factor could be to multiply the two factors together and then divide the answer into the polynomial. There might be an easier way.

Starting the solution

1   Let $f(x) = 2x^3 + px^2 + qx - 12$ and as $f(2)$ and $f(-2)$ both equal zero, we can obtain the two simultaneous equations to solve to find the values of p and q.

2   We can let the unknown factor be $(ax + b)$ and then multiply this by the other two factors and then equate the relevant coefficients to find the value of a and hence the factor.

The solution

(a) $f(x) = 2x^3 + px^2 + qx - 12$

$f(2) = 16 + 4p + 2q - 12$ and as $(x - 2)$ is a factor, $f(2) = 0$, so $4p + 2q + 4 = 0$.

$f(-2) = -16 + 4p - 2q - 12$ and as $(x + 2)$ is a factor, $f(-2) = 0$, so $4p - 2q - 28 = 0$.

> The two equations are now solved simultaneously.

Adding these two equations together

$8p - 24 = 0$ giving $p = 3$

Substituting into the first equation gives $12 + 2q + 4 = 0$ so $q = -8$

(b) $(x + 2)(x - 2)(ax + b) = (x^2 - 4)(ax + b)$

$$= ax^3 + bx^2 - 4ax - 4b$$

> Note that we need a number, a, in front of the x for the unknown factor $(ax + b)$.

Now, $ax^3 + bx^2 - 4ax - 4b = 2x^3 + 3x^2 - 8x - 12$

Equating coefficients of x^3, we obtain $a = 2$.

Equating coefficients independent of x, we obtain $b = 3$.

Hence the factor is $(2x + 3)$

---

5   Prove that there is only **one** real root of the equation
$8x^3 + 7x^2 - 13x + 10 = 0$.   [7]

Thinking about the question

We need to first find the real root and then the quadratic that this multiplies to give the equation given.

Starting the solution

Let $f(x) = 8x^3 + 7x^2 - 13x + 10$ and then substitute values in for x until the function equals zero. We then find the factor.

We can then find the quadratic function and see if the discriminant is less than 0, which is the condition for no real roots.

The solution

$$f(x) = 8x^3 + 7x^2 - 13x + 10$$

Let $x = 1$, so $f(1) = 8(1)^3 + 7(1)^2 - 13(1) + 10 = 12$

Let $x = -1$, so $f(-1) = 8(-1)^3 + 7(-1)^2 - 13(-1) + 10 = 22$

Let $x = 2$, so $f(2) = 8(2)^3 + 7(2)^2 - 13(2) + 10 = 76$

Let $x = -2$, so $f(-2) = 8(-2)^3 + 7(-2)^2 - 13(-2) + 10 = 0$

Hence $(x + 2)$ is a factor of the equation.

$(x + 2)(ax^2 + bx + c) = 8x^3 + 7x^2 - 13x + 10$

Equating coefficients of x^3 we obtain $a = 8$

Equating coefficients independent of x we obtain $2c = 10$ so $c = 5$.

Equating coefficients of x^2 we obtain $b + 2a = 7$ so $b + 16 = 7$, $b = -9$.

Hence we have $(x + 2)(8x^2 - 9x + 5)$

For $8x^2 - 9x + 5$, $b^2 - 4ac = 81 - 160 = -79$

$b^2 - 4ac < 0$, so the quadratic factor has no real roots.

The only real root is $(x + 2)$.

> To prove that $8x^2 - 9x + 5$ has no real roots we need to find the discriminant and show that it is less than zero.

---

6   The diagram below shows a sketch of $y = f(x)$.

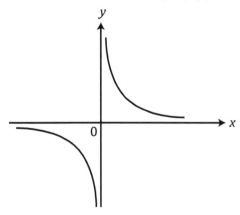

(a)   Sketch the graph of $y = 4 + f(x)$, clearly indicating any asymptotes.   [2]

(b)   Sketch the graph of $y = f(x - 3)$, clearly indicating any asymptotes.   [2]

Thinking about the question

You should recognise the curve is for an equation in the form $y = \frac{a}{x}$ and this is a question on transformations.

Starting the solution

There are two asymptotes to the curve: the x-axis and the y-axis. The position of these asymptotes will change depending on the transformation.

> An asymptote is a line which a curve approaches, but never actually touches.

The solution

(a) $y = 4 + f(x)$ is a translation by $\binom{0}{4}$ so the curve is shifted up 4 units parallel to the positive y-direction.

The asymptotes will now be at $x = 0$ and $y = 4$.

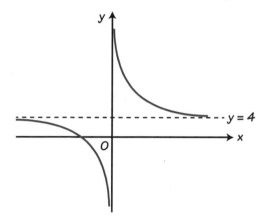

> Make sure that you clearly label any asymptotes other than the axes and write their equations.

(b) $y = f(x - 3)$ is a translation by $\binom{3}{0}$ so the curve is shifted 3 units parallel to the positive x-direction.

The asymptotes will now be at $x = 3$ and $y = 0$.

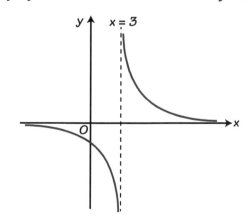

> **Common mistakes**
> You must retain the asymptotes when drawing the transformed curve. Make sure when drawing the curve that it does not touch, or appear to be going to touch, any of the asymptotes.

Exam practice

① Show that $(\sqrt{5} - \sqrt{3})^3$ can be written as $2(a\sqrt{5} - b\sqrt{3})$ where a and b are integers. [4]

② (a) Show that $\dfrac{2\sqrt{5} + \sqrt{3}}{\sqrt{5} + \sqrt{3}}$ can be written as $\dfrac{1}{2}(7 - \sqrt{15})$. [3]

(b) If $(3^a)^3 \times 3^a \times 3 = 81$, find the value of a. [3]

③ Simplify

(a) $\dfrac{10}{7 + 2\sqrt{11}}$ [3]

(b) $(4\sqrt{3})^2 - (\sqrt{8} \times \sqrt{50}) - \dfrac{5\sqrt{63}}{\sqrt{7}}$ [4]

④ The cubic polynomial $f(x)$ is given by $f(x) = 2x^3 + ax^2 + bx + c$, where a, b, c are constants. The graph of $f(x)$ intersects the x-axis at the points with coordinates $(-3, 0)$, $(2.5, 0)$ and $(4, 0)$. Find the coordinates of the point where the graph of $f(x)$ intersects the y-axis. [5]

⑤ The quadratic equation $4x^2 - 12x + m = 0$, where m is a positive constant, has two distinct real roots. Show that the quadratic equation $3x^2 + mx + 7 = 0$ has no real roots. [4]

⑥ The diagram shows a sketch of the graph of $y = f(x)$. The graph passes through the points $(-4, 0)$ and $(6, 0)$ and has a maximum point at $(1, 3)$.

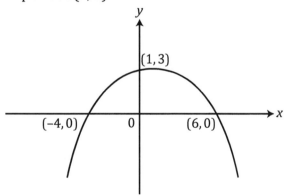

(a) Sketch the graph of $y = f(x + 3)$, indicating the coordinates of the stationary point and the coordinates of the points of intersection of the graph with the x-axis. [3]

(b) Gwen is asked by her teacher to draw the graph of $y = f(ax)$ for various values of the constant a. Two of Gwen's graphs pass through the point $(2, 0)$. Find the value of a corresponding to each of these two graphs. [2]

7 (a) Factorise $2x^3 + 3x^2 - 3x - 2$. [4]
   (b) Find the remainder when $2x^3 + 3x^2 - 3x - 2$ is divided by
       $2x - 1$. [2]

8 Find the greatest value of $\dfrac{1}{3x^2 - 12x + 15}$. [6]

9 (a) Express $4x^2 + 40x - 69$ in the form $a(x + b)^2 + c$, where the
       values of the constants a, b and c are to be found. [3]
   (b) Using your answer to part (a), solve the equation
       $4x^2 + 40x - 69 = 0$. [3]

10 Simplify $\sqrt{500} + \left(\sqrt{12} \times \sqrt{15}\right) - \dfrac{7\sqrt{60}}{\sqrt{3}}$ [4]

> We need to first complete the square to find the minimum value of $3x^2 - 12x + 15$. One divided by this value will give the greatest value.

3 Coordinate geometry in the (x, y) plane

Prior knowledge

You will need to make sure you fully understand the following from your GCSE studies:

- Using equations and formulae
- Solving simultaneous equations
- Angle and tangent properties of circles

Quick revision

The gradient of the line joining points (x_1, y_1) and (x_2, y_2) is given by:

$$\text{Gradient} = \frac{y_2 - y_1}{x_2 - x_1}$$

You need to remember this formula as it will not be given in the formula booklet.

The length of a straight line joining the two points (x_1, y_1) and (x_2, y_2) is given by:

$$\sqrt{(x_2 - x_1)^2 + (y_2 - y_1)^2}$$

Remember this formula.

The mid-point of a line joining the points (x_1, y_1) and (x_2, y_2) is given by:

$$\left(\frac{x_1 + x_2}{2}, \frac{y_1 + y_2}{2} \right)$$

Remember this formula.

The equation of a straight line with gradient m and which passes through a point (x_1, y_1) is given by:

$$y - y_1 = m(x - x_1)$$

Remember this formula.

Condition for two straight lines to be parallel to each other – both lines must have the same gradient.

Condition for two straight lines to be perpendicular to each other – the product of their gradients is -1. If one line has a gradient m_1 and the other a gradient of m_2 then $m_1 m_2 = -1$

The equation of a circle

The equation of a circle with centre (a, b) and radius r is:

$$(x - a)^2 + (y - b)^2 = r^2$$

There is the following alternative form for the equation of a circle:

$$x^2 + y^2 + 2gx + 2fy + c = 0$$

None of the formulae for circles are included in the formula booklet.

and a circle having this equation has centre $(-g, -f)$ and radius given by

$$\sqrt{g^2 + f^2 - c}$$

Circle properties

There are a number of circle properties you need to know about:

1   The angle in a semicircle is always a right angle.

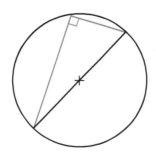

2 The perpendicular from the centre of the circle to a chord bisects the chord.

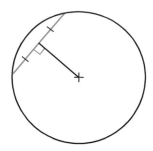

3 The tangent to a circle at a point makes a right angle to the radius of the circle at the same point.

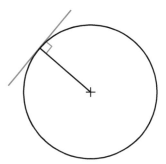

Finding where a circle and straight line intersect or meet

A line and a circle can meet in either one place (i.e. the line is a tangent to the circle) or intersect in two places.

1 Solve the two equations (i.e. the circle and the line) simultaneously by equating the y-values.

2 Obtain a quadratic equation and solve it. Two roots mean the line intersects the circle in two places. Two equal roots mean there is only one point so the line is a tangent to the circle.

Proving the line and circle do not intersect or meet – obtain the quadratic equation as above but find the discriminant and if $b^2 - 4ac < 0$ then there are no real roots so the circle and line do not intersect or meet.

Condition for two circles to touch internally – the distance between the centres of the circles must equal the difference in the radii of the two circles.

Condition for two circles to touch externally – the distance between the two centres of the circles must equal the sum of the two radii of the circles.

Looking at exam questions

1   The points A, B, C have coordinates (4, −2), (−12, 10), (10, 6), respectively.

(a)   Find the gradients of the lines AB, BC, CA.   [3]

(b)   Show that one of the angles of triangle ABC is a right angle.   [2]

(c)   Show that the equation of the line AB is

$$3x + 4y − 4 = 0.$$ [2]

(d)   The mid-point of BC is D. Find the length of AD.   [4]

Thinking about the question

There are a few points to deal with so we need to draw a graph. We can use the graph to check the signs of gradients, etc. You can add the various points to the diagram as you proceed through the question.

Starting the solution

(a) Draw a set of axes and add the points A, B and C.

We can use the equation for the gradient of a line joining two points to find the required gradients. We can refer to the graph to check that we have the correct sign for the gradient of each line.

(b) It should be possible to see from the graph which lines are at right angles; however, this alone should not be used, as you need to check that the product of the two lines making the suspected right angle multiply together to give −1.

(c) In order to find the equation of a line, you need to know the gradient, which was found in part (a) and a point through which the line passes (i.e. either point A or B in this case). The equation $y − y_1 = m(x − x_1)$ can then be used.

(d) The coordinates of B and C are used with the formula

$$\left(\frac{x_1 + x_2}{2}, \frac{y_1 + y_2}{2}\right)$$

to find the mid-point D. The formula for the distance between two points can be used with the coordinates of A and D to find the length of AD.

The solution

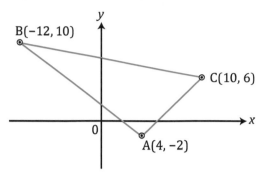

Watch out

It is easy to make mistakes when substituting coordinates into the formula for the gradient. It is a good idea to check the sign of your gradients with your graph.

(a) Gradient of AB $= \dfrac{10 - (-2)}{-12 - 4} = \dfrac{12}{-16} = -\dfrac{3}{4}$

Gradient of BC $= \dfrac{10 - 6}{-12 - 10} = \dfrac{4}{-22} = -\dfrac{2}{11}$

Gradient of CA $= \dfrac{6 - (-2)}{10 - 4} = \dfrac{8}{6} = \dfrac{4}{3}$

(b) From the diagram we see that lines AB and CA are likely to be at right angles to each other.

$$m_1 m_2 = -1$$

$$\left(-\dfrac{3}{4}\right)\left(\dfrac{4}{3}\right) = -1 \quad \text{so lines AB and CA are perpendicular.}$$

(c) Equation of AB is:

$$y - (-2) = -\dfrac{3}{4}(x - 4)$$

$$y + 2 = -\dfrac{3}{4}(x - 4)$$

$$3x + 4y - 4 = 0$$

(d) Mid-point, D $= \left(\dfrac{x_1 + x_2}{2}, \dfrac{y_1 + y_2}{2}\right) = \left(\dfrac{-12 + 10}{2}, \dfrac{10 + 6}{2}\right) = (-1, 8)$

The length of a straight line joining the two points (x_1, y_1) and (x_2, y_2) is given by:

$$d = \sqrt{(x_2 - x_1)^2 + (y_2 - y_1)^2}$$

Now A is (4, −2) and D is (−1, 8)

$$AD = \sqrt{(-1 - 4)^2 + (8 - -2)^2} = \sqrt{125} = 5\sqrt{5}$$

---

2 The points A and B have coordinates (−1, 10) and (5, 1) respectively. The straight line L has equation $2x - 3y + 6 = 0$.

(a) The line L intersects the line AB at the point C. Find the coordinates of C. [5]

(b) Determine the ratio in which the line L divides the line AB. [2]

(c) The line L crosses the *x*-axis at the point D.
Find the coordinates of D. [1]

(d) (i) Show that L is perpendicular to AB.

(ii) Calculate the area of the triangle ACD. [6]

Thinking about the question

1 Look at the mark allocation which gives you an idea of the complexity and the time that should be given to the question. The number of marks allocated is high (in fact it was the highest mark allocation for a question on the whole paper). Time needs to be spent on this question.

2 Read the whole of the question before you start. By reading the whole question you can see how one part leads to another and this can give an indication of how to solve a particular part you are not sure of.

3 As it involves coordinates, always spend time drawing diagrams.

Starting the solution

(a) Here we are looking for the point where a line whose equation is given, intersects a line passing through two points.

To find this point of intersection, we need to find the gradient of the line joining the points and hence its equation. Both equations can then be solved simultaneously to find the coordinates of the point of intersection.

(b) You make think that as this part involves the lengths of lines, you use the formula for the distance between two points to find the lengths of the two line segments.

There is an easier method which is explained in the solution.

(c) Here you simply substitute the equation of the *x*-axis (i.e. $y = 0$) into the equation of the line L to find the *x*-coordinate. The *y*-coordinate will be 0.

(d) (i) Notice here you have to prove two lines are perpendicular. To do this, find the gradient of each line and if they are perpendicular, when the gradients are multiplied together it will give −1.

> Where there are two parts like (i) and (ii), usually you find the answer to part (i) is used in part (ii).

(ii) Ask yourself why you have proved that the two lines are perpendicular. It is likely that the triangle is right angled, which means that the area can be found simply. The base and height of the triangle can be found using the formula for the distance between two points.

The solution

The formula for a straight line is used here.

The formula for the equation of a straight line having gradient m and passing through the point (x_1, y_1) is $y - y_1 = m(x - x_1)$. Note that this formula must be remembered.

(a)   Gradient of $AB = \dfrac{1 - 10}{5 - (-1)} = \dfrac{-9}{6} = -\dfrac{3}{2}$

Equation of AB is $y - 1 = -\dfrac{3}{2}\left(x - 5\right)$

$$2y - 2 = -3x + 15$$

$$3x + 2y - 17 = 0$$

Solving this equation simultaneously with the equation of L:

We have the two equations $3x + 2y - 17 = 0$ and $2x - 3y + 6 = 0$

Multiplying the first equation by 3 and the second equation by 2 to make the coefficients $6y$ and $-6y$ so that the equations can be added to eliminate the term in y.

$$9x + 6y - 51 = 0$$

$$4x - 6y + 12 = 0$$

Adding, we obtain $13x - 39 = 0$, giving $x = 3$.

Substituting this value into either equation, we obtain $y = 4$.

Hence C is the point $(3, 4)$.

(b)

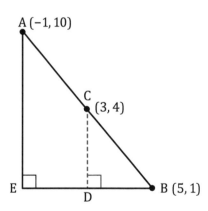

Watch out

Make it clear which sides you are referring to in your ratio. Don't just write 'ratio is 2 : 1'. You need to say which side is 2 and which side is 1.

There are other ways to find this ratio. You could have used the formula for the distance between two points to find the lengths.

The ratio of AC : CB will be the same as the ratio of DE : BD

Hence   DE : BD = $3 - -1 : 5 - 3$

$$= 4 : 2$$

$$= 2 : 1$$

So   AC : CB = 2 : 1

(c) The equation of the x-axis is $y = 0$ so substituting this into the equation for L we obtain $2x - 3(0) + 6 = 0$ giving $x = -3$.

D is the point $(-3, 0)$

> Make sure you give the coordinates of D and don't just give the x-value.

(d) (i) Rearranging $2x - 3y + 6 = 0$ so that it is in the form $y = mx + c$, gives

$$y = \frac{2}{3}(x + 2)$$

Gradient of L is $\frac{2}{3}$

From part (a), gradient of AB $= -\frac{3}{2}$

Now, $\left(\frac{2}{3}\right)\left(-\frac{3}{2}\right) = -1$ which proves these two lines are perpendicular.

> Remember that if two lines are perpendicular, the product of their gradients is -1.

(ii) As line L is perpendicular to AB and C is the point on AB where L cuts AB, it means that AC and AD are at right angles so triangle ACD is a right-angled triangle.

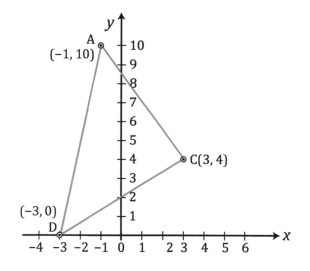

$$AC = \sqrt{(10 - 4)^2 + (-1 - 3)^2} = \sqrt{52}$$

$$CD = \sqrt{(4 - 0)^2 + (3 - -3)^2} = \sqrt{52}$$

Area of triangle ACD $= \frac{1}{2} \times \sqrt{52} \times \sqrt{52}$

$$= 26 \text{ units}^2$$

3   A circle C has equation

$$x^2 + y^2 - 10x - 14y + 49 = 0$$

(a) Find the equation of the tangent to the circle C at the point (2, 3). [6]

(b) (i) Show that $Q(13, 13)$ lies outside the circle C. [2]

(ii) Find the equation of a circle with centre at Q which touches the circle C externally. [3]

Thinking about the question

The mark allocation for part (a) is quite high and there are no pointers as to what you have to do. For part (a) you need to first find the coordinates of the centre and the radius of the circle.

For part (b)(i) we need to find the length of the line joining the centre of the circle to Q. If this line is longer than the radius, then Q lies outside the circle.

For part (b)(ii) we need to use the fact that the distance between the two centres of the circles will equal the sum of the two radii of each circle in order to find the equation of the other circle. If you know the centre and the radius of a circle, then its equation can be found.

Starting the solution

For part (a) we find the coordinates of the centre and the radius. We can then use the coordinates of the centre and the point (2, 3) to find the gradient of the radius joining these two points. Then we can find the gradient of the tangent using $m_1m_2 = -1$ and then use the formula $y - y_1 = m(x - x_1)$ to find the equation of the tangent.

For (b)(i) we can find the distance from the point Q to the centre of the circle and if it is greater than the radius then point Q lies outside the circle.

For (b)(ii) we use the fact that the distance between the two centres must equal the sum of the radii for the circles to touch externally.

The solution

(a) $$x^2 + y^2 - 10x - 14y + 49 = 0$$

$$(x - 5)^2 + (y - 7)^2 - 25 - 49 + 49 = 0$$

Hence $$(x - 5)^2 + (y - 7)^2 = 25$$

So the coordinates of the centre of the circle are (5, 7) and the radius is 5.

Here we will complete the square but you could also use the formula:

$$x^2 + y^2 + 2gx + 2fy + c = 0$$

to find the centre $(-g, -f)$ and radius given by:

$$\sqrt{g^2 + f^2 - c}$$

Gradient of the radius joining the centre (5, 7) to the point (2, 3) $= \frac{7-3}{5-2} = \frac{4}{3}$

Gradient of the tangent $= -\frac{3}{4}$.

Equation of tangent, $\quad y - 3 = -\frac{3}{4}(x - 2)$

$$4y - 12 = -3x + 6$$

$$3x + 4y - 18 = 0$$

> Here we use the fact that the radius and tangent are at right angles to each other, so the product of their gradients is –1.

(b) (i) Distance from the centre of the circle (5, 7) and point Q (13, 13)

$$= \sqrt{(x_2 - x_1)^2 + (y_2 - y_1)^2}$$

$$= \sqrt{(13 - 5)^2 + (13 - 7)^2}$$

$$= 10$$

> If you have difficulty remembering this formula you could alternatively plot the two points on the graph and form a right-angled triangle and use Pythagoras' theorem.

The radius of the circle is 5, so this point lies outside the circle.

(ii) For both circles to touch externally, the distance between the centres must equal the sum of the radii of the two circles. Now the distance between the centres, i.e. (5, 7) and (13, 13), is 10 and since the radius of C is 5, the radius of the circle with centre Q must also be 5.

Hence equation of circle is $(x - 13)^2 + (y - 13)^2 = 25$.

Exam practice

1 (a) Find the centre and radius of the circle C given by
$$x^2 + y^2 - 8x + 4y + 11 = 0.$$ [3]

(b) Given that the circle $x^2 + y^2 = a^2$ $(a > 0)$ touches C externally, find the value of a, giving your answer correct to two decimal places. [4]

2 Circle C has equation $x^2 + y^2 - 8x + 6y - 24 = 0$.
(a) Find the coordinates of the centre of C. [3]
(b) Show that the point P(−4, 3) lies outside the circle. [2]

3 (a) Show that the circles C_1 and C_2 having equations
$$x^2 + y^2 - 2x + 8y - 8 = 0 \quad \text{and} \quad x^2 + y^2 - 2x + 8y - 19 = 0$$
respectively, share the same centre. [4]

(b) (i) Prove the point P(4, 0) lies on circle C_1. [1]
(ii) Find the equation of the tangent to the circle C_1 at point P. [2]

④ The points A, B, C have coordinates (−2, −3), (6, 1) and (k, 3) respectively. The line AB is perpendicular to BC.
(a) Find the gradient of AB.   [2]
(b) Show that $k = 5$.   [3]
(c) The line L is parallel to BC and passes through A.
      Find the equation of L.   [2]
(d) The line L intersects the y-axis at D. Calculate the
      length of CD.   [3]

⑤ Points A (2, 1) and B (0, −5) are opposite ends of a diameter of circle C. Find the equation of circle C.   [5]

⑥ The circle shown below has centre (−4, 3) and it intersects the x-axis at the origin O and point B.

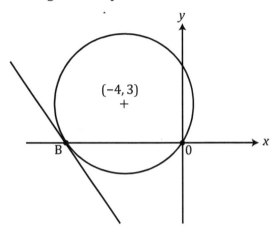

Find
(a) The equation of the tangent to the circle at point B.   [3]
(b) The coordinates of the point where the tangent at B
      intersects the y-axis.   [5]

⑦ The circle C has radius 5 and its centre is the origin.
The point T has coordinates (11, 0).
The tangents from T to the circle C touch C at the points R and S.
(a) Write down the geometrical name for the quadrilateral
      ORTS.   [1]
(b) Find the exact value of the area of the quadrilateral ORTS.
      Give your answer in its simplest form.   [5]

4 Sequences and series – the binomial theorem

Prior knowledge

You will need to make sure you fully understand the following from your GCSE studies:

- The use of index form
- Substituting numbers into formulae

Quick revision

The binomial expansion

Binomial expansion is the expansion of the expression $(a + b)^n$ where n is a positive integer.

The formula for the expansion will be given in the formula booklet and is shown here:

$$(a + b)^n = a^n + \binom{n}{1}a^{n-1}b + \binom{n}{2}a^{n-2}b^2 + \ldots + \binom{n}{r}a^{n-r}b^r + \ldots + b^n$$

where $\binom{n}{r} = {}^nC_r = \dfrac{n!}{r!(n-r)!}$

> You do not need to memorise these formulae as they are given in the formula booklet.

$n!$ means n factorial. If $n = 5$ then $5! = 5 \times 4 \times 3 \times 2 \times 1$

Note that $0! = 1$.

Pascal's triangle

> You won't be given Pascal's triangle in the formula booklet.
>
> Notice that all the rows start and end with a 1. Notice also that the other numbers are found by adding the pairs of numbers immediately above. For example, if we have 1   3 in the line above then the number to be entered between these numbers on the next line is a 4.

```
            1
          1   1
        1   2   1
      1   3   3   1
    1   4   6   4   1
  1   5   10  10  5   1
```

It can be used to find the coefficients in the expansion of $(a + b)^n$.

If you want the coefficients when $n = 3$, look for the line with 1 and then 3. The coefficients are therefore 1   3   3   1.

The binomial expansion when $a = 1$

> Again this formula is given in the formula booklet so you don't need to memorise it.

$$(1 + x)^n = 1 + nx + \frac{n(n-1)}{2!}x^2 + \frac{n(n-1)(n-2)}{3!}x^3 + \ldots$$

Looking at exam questions

1 (a) Use the binomial theorem to expand $\left(a + \sqrt{b}\right)^4$.   [2]

    (b) Hence, deduce an expression in terms of a and b for
$$\left(a + \sqrt{b}\right)^4 + \left(a - \sqrt{b}\right)^4$$ [2]

Thinking about the question

This is a question involving the binomial theorem. The formula can be obtained from the formula sheet. We note that the answer to part (a) is used in part (b).

Starting the solution

For part (a) we can use the formula for the expansion of $(a + b)^n$ with $n = 4$, $a = a$ and $b = \sqrt{b}$. We can use Pascal's triangle for the coefficients or alternatively use a calculator.

For part (b) we use our answer to part (a) for the expansion of $\left(a + \sqrt{b}\right)^4$ and for $\left(a - \sqrt{b}\right)^4$ we need to adapt the expansion when we have $-\sqrt{b}$ instead of $+\sqrt{b}$. We need to be careful of the signs here.

The solution

(a) $(a + b)^n = a^n + \binom{n}{1}a^{n-1}b + \binom{n}{2}a^{n-2}b^2 + \ldots + \binom{n}{r}a^{n-r}b^r + \ldots + b^n$

For $\left(a + \sqrt{b}\right)^4$, $n = 4$, so using Pascal's triangle for the coefficients, we look for the line in the triangle starting with 1 and then 4. Here is the line:

$$1 \quad 4 \quad 6 \quad 4 \quad 1$$
$$\left(a + \sqrt{b}\right)^4 = a^4 + 4a^3\sqrt{b} + 6a^2b + 4ab\sqrt{b} + b^2$$

(b) $\left(a + \sqrt{b}\right)^4 + \left(a - \sqrt{b}\right)^4$

$$= a^4 + 4a^3\sqrt{b} + 6a^2b + 4ab\sqrt{b} + b^2$$
$$+ \left(a^4 - 4a^3\sqrt{b} + 6a^2b - 4ab\sqrt{b} + b^2\right)$$

$$= a^4 + 4a^3\sqrt{b} + 6a^2b + 4ab\sqrt{b} + b^2 + a^4 - 4a^3\sqrt{b} + 6a^2b - 4ab\sqrt{b} + b^2$$

$$= 2a^4 + 12a^2b + 2b^2$$

---

2 If $(1 + ax)^n = 1 + 10x + 60x^2 + \ldots$

Find the value of n and the value of a. [5]

Thinking about the question

This is a question on the binomial expansion, so we need to look up the formula for the expansion of $(1 + x)^n$ in the formula booklet.

Starting the solution

Expand $(1 + ax)^n$ as far as the term in x^2. We can then compare this expansion to the one given in the question. We then compare coefficients of x and then x^2 in order to find the values of a and n.

The solution

$$(1 + x)^n = 1 + nx + \frac{n(n-1)}{2!}x^2 + \ldots$$

$$(1 + ax)^n = 1 + anx + \frac{n(n-1)}{2!}(ax)^2 + \ldots$$

> Here you replace x in the formula with ax.

Equating 2nd terms, $\qquad anx = 10x$, so $an = 10$

Equating 3rd terms, $\dfrac{n(n-1)}{2!}(ax)^2 = 60x^2$

$$n(n-1)a^2 = 120$$

Now $\qquad a = \dfrac{10}{n}$

So, $\quad n(n-1)\dfrac{100}{n^2} = 120$

$$100n^2 - 100n = 120n^2$$

$$20n^2 + 100n = 0$$

$$20n(n+5) = 0$$

$n = 0$ or -5

> **Watch out**
>
> Don't be tempted to divide both sides by n. If you do this, you will lose one of the solutions.

We can reject 0 as if this is substituted into the series, we would only obtain 1 term (i.e. 1).

Hence, $n = -5$

> Whenever you obtain two answers to a question, always ask yourself whether both of them are allowed or just one of them. Here 0 is not allowed as you would not obtain more than one term.

Now $a = \dfrac{10}{n} = \dfrac{10}{-5} = -2$

So, $n = -5$ and $a = -2$

Exam practice

1. (a) Write down the expansion of $(a + b)^4$. [2]
   (b) In the binomial expansion of $(a + 2x)^4$, the coefficient of the term in x^2 is twelve times the coefficient of the term in x^3. Find the value of a. [3]

2. (a) Write down the expansion of $(1 + x)^6$ in ascending powers of x up to and including the term in x^3. [2]
   (b) By substituting an appropriate value for x in your expansion in (a), find an approximate value for 0.99^6. Show all your working and give your answer correct to four decimal places. [3]

3. Find the term in x^2 in the binomial expansion of $\left(x + \dfrac{3}{x}\right)^6$. [4]

4. (a) Expand $(2x + 3)^4$, simplifying each term of the expansion. [3]
   (b) In the binomial expansion of $(1 + 3x)^n$ the coefficient of x^2 is 54. Given that $n > 0$, find the value of n. [3]

5. Using the Binomial theorem, fully expand $\left(1 + \sqrt{x}\right)^4$ [4]

5 Trigonometry

Prior knowledge

You will need to make sure you fully understand the following from your GCSE studies:

- Pythagoras' theorem
- Trigonometric relationships in right-angled triangles
- Sine and cosine rules
- The graphs of trigonometric functions
- Using the formula: Area of a triangle = $\frac{1}{2}ab\sin C$

Quick revision

Sine, cosine and tangent functions and their exact values

$$\sin 30° = \frac{1}{2} \qquad \sin 45° = \frac{1}{\sqrt{2}} \qquad \sin 60° = \frac{\sqrt{3}}{2}$$

$$\cos 30° = \frac{\sqrt{3}}{2} \qquad \cos 45° = \frac{1}{\sqrt{2}} \qquad \cos 60° = \frac{1}{2}$$

$$\tan 30° = \frac{1}{\sqrt{3}} \qquad \tan 45° = 1 \qquad \tan 60° = \sqrt{3}$$

Obtaining angles given a trigonometric ratio

There are two methods:

Using the CAST method
This uses this diagram:

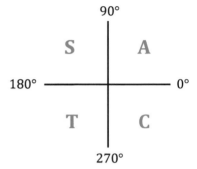

> Letter A means all the trig ratios are positive, S means only sine is positive, T means only tan is positive and C means only cos is positive.

Suppose you want to find the values of θ in the range $0° \le \theta \le 360°$ where $\sin \theta = \frac{1}{2}$.

The value $\frac{1}{2}$ is positive, and sin is positive in the first and second quadrants, so we can show this as follows:

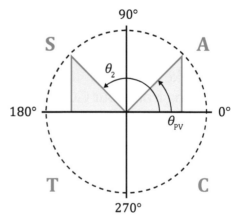

To find θ_{pv} we use $\theta_{pv} = \sin^{-1}\left(\frac{1}{2}\right)$, giving $\theta_{pv} = 30°$.
As the shaded triangles are the same, θ_2 can be found by subtracting 30° from 180° to give $\theta_2 = 150°$. The two values are 30° and 150°.

Using the trigonometric graphs

You draw the relevant graph in the range specified (i.e. sin, cos or tan). Draw a horizontal line corresponding to the trig ratio. For example, if $\cos x = -\frac{1}{2}$, you would draw a horizontal line crossing the y-axis at $y = -\frac{1}{2}$ and then read off the corresponding values on the x-axis. You may have to use the symmetry of the graphs to obtain the angles.

The Sine and Cosine rules and the formula for the area of a triangle

The Sine rule states: $\quad \dfrac{a}{\sin A} = \dfrac{b}{\sin B} = \dfrac{c}{\sin C}$

All these formulae will need to be remembered.

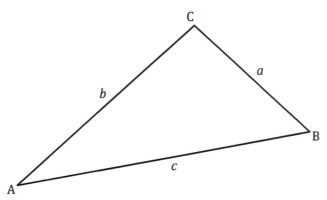

The Cosine rule states: $\quad a^2 = b^2 + c^2 - 2bc \cos A$

$$\text{Area of triangle} = \frac{1}{2}ab \sin C$$

Trigonometric relationships

$$\cos^2 \theta + \sin^2 \theta = 1$$

$$\tan \theta = \frac{\sin \theta}{\cos \theta}$$

Radian measurement

$$1 \text{ radian} = \frac{180}{\pi} = \frac{180}{3.14} = 57.3°$$

Here are some popular angles expressed in radians and degrees:

2π radians = 360° π radians = 180°

$\dfrac{\pi}{2}$ radians = 90° $\dfrac{\pi}{4}$ radians = 45°

$\dfrac{\pi}{3}$ radians = 60° $\dfrac{\pi}{6}$ radians = 30°

Watch out

Be careful when using this formula when working out the angle when the area and two sides of the triangle are known. For example, $\sin C = \frac{1}{2}$ can have two solutions 30° and 150°. If the triangle has been drawn, or other angles in the triangle are known then the obtuse angle might not be a possible solution. Be guided by the wording of the question, so look for the plural 'angles' in the question to see if you are looking for two angles.

Watch out

Always change your calculator back to degrees after you have completed radian calculations.

Sine, cosine and tangent: their graph, symmetries and periodicity

The sine graph ($y = \sin \theta$) where θ is expressed in radians

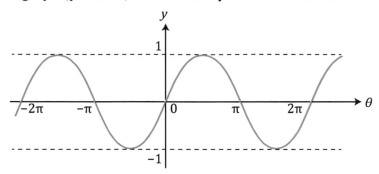

The sine graph has a period of 2π meaning the graph repeats itself every 2π radians.

The sine graph has a period of 2π as a particular value of θ will have the same y-value as an angle of $\theta + 2\pi$, $\theta + 4\pi$, and so on.

The sine graph ($y = \sin \theta$) where θ is expressed in degrees

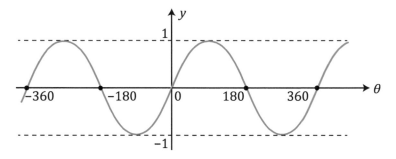

The cosine graph ($y = \cos \theta$) where θ is expressed in radians

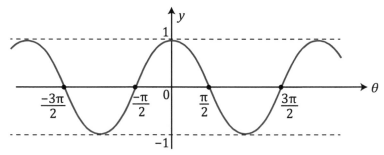

The cosine graph has a period of 2π meaning the graph repeats itself every 2π radians.

The cosine graph has a period of 2π as a particular value of θ will have the same y-value as an angle of $\theta + 2\pi$, $\theta + 4\pi$, and so on.

The cosine graph ($y = \cos \theta$) where θ is expressed in degrees

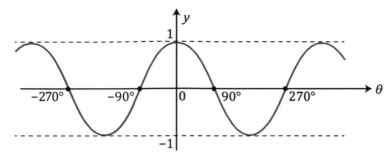

The tangent graph ($y = \tan \theta$) where θ is expressed in radians

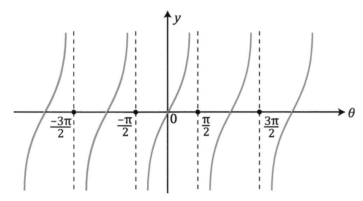

The period of the tan θ graph is π radians.

The tangent graph ($y = \tan \theta$) where θ is expressed in degrees

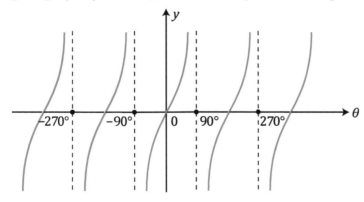

Looking at exam questions

1   The triangle ABC is such that AC = 16 cm, AB = 25 cm and ABC = 32°.
Find two possible values for the area of the triangle ABC.   [5]

Thinking about the question

A diagram of the triangle must be drawn, marking on it the known lengths and angle.

If it is not a right-angled triangle then we probably have to find the area of the triangle using two sides and the included angle with

$$\text{Area} = \frac{1}{2}ab \sin C.$$

Starting the solution

When drawing the diagram try to draw it in proportion and get the angles approximately correct but it is not necessary to start producing an accurate diagram using a ruler and protractor.

We also need to think how it is possible for a triangle to have two different areas. One of the angles must be either acute or obtuse so we need to be careful when solving to find the angles.

The solution

The angle α between the two sides needs to be found so that the formula Area of triangle = $\frac{1}{2}ab \sin C$ can be used to find the areas. This formula is on the formulae sheet.

Angle α is hard to find directly.

Notice that it is easy to find θ first using the Sine rule and then use this with 32° to find angle α.

The formula for the Sine rule is included on the formula sheet.

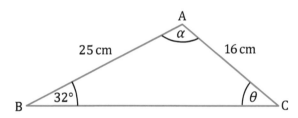

Using the Sine rule, we have $\dfrac{25}{\sin \theta} = \dfrac{16}{\sin 32°}$

Solving, gives $\theta = 55.8937°$ or $180° - 55.8937° = 124.1063°$

Now $\alpha = 180 - (\theta + 32)$

Hence $\alpha = 92.1063°$ or $23.8937°$

Using $\alpha = 92.1063°$, Area = $\frac{1}{2} \times 25 \times 16 \sin 92.1063° = 199.86$ cm²

Using $\alpha = 23.8937°$, Area = $\frac{1}{2} \times 25 \times 16 \sin 23.8937° = 81.01$ cm²

So area of triangle ABC = 199.86 cm² or 81.01 cm²

2   Solve the following equation for values of θ between $0°$ and $360°$.

$$2 - 3\cos^2\theta = 2\sin\theta. \qquad\qquad [6]$$

Thinking about the question

This is a trigonometric equation but as it involves both sin and cos we need to form a quadratic equation in either sin or cos.

Starting the solution

It is easier to change from $\cos^2\theta$ to $\sin^2\theta$ so this is what we do. We can then solve the resulting quadratic equation.

The solution

> $\sin^2\theta + \cos^2\theta = 1$
> so $\cos^2\theta = 1 - \sin^2\theta$

$$2 - 3\cos^2\theta = 2\sin\theta.$$

Hence $\qquad 2 - 3(1 - \sin^2\theta) = 2\sin\theta$

$$2 - 3 + 3\sin^2\theta = 2\sin\theta$$

$$3\sin^2\theta - 2\sin\theta - 1 = 0$$

$$(3\sin\theta + 1)(\sin\theta - 1) = 0$$

Hence $\sin\theta = -\frac{1}{3}$ or $\sin\theta = 1$

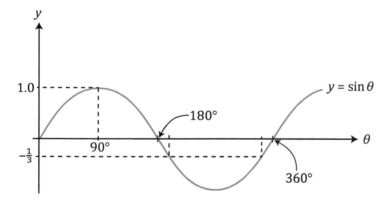

$\sin\theta = 1$, gives $\theta = 90°$

$\sin\theta = -\frac{1}{3}$, gives $\theta = 199.47°, 340.53°$

Hence $\theta = 90°, 199.47°, 340.53°$

Exam practice

1 The triangle ABC is such that AB = x cm, BC = $(x - 3)$ cm,
CA = $(x - 1)$ cm and angle ABC = 60°.
(a) Use the cosine rule to show that $x = 8$. [4]
(b) Find the area of triangle ABC, giving your answer in
surd form. [2]

2 Solve the trigonometric equation: $1 - \tan 2x = 3$
for values of x between 0° and 180°. [4]

3 The triangle ABC is such that AB = 12 cm, BC = 10 cm and $C\hat{A}B = 45°$.
(a) Find, to the nearest degree, the two possible values of $B\hat{C}A$. [3]
(b) Find, correct to one decimal place, the possible values of
the length AC. [5]

4 The diagram shows triangle ABC, with area 12 cm², AB = 5 cm
and AC = 8 cm.

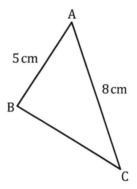

Given that $B\hat{A}C$ is an acute angle, find the length of BC correct to
one decimal place. [3]

5 Find the values of x in the range $0° \le x \le 360°$, that satisfy the
equation: $3 \sin x = \tan x$ [4]

6 Find the values of x in the range $0° \le x \le 360°$, that satisfy the
equation: $6 \sin^2 x = 4 + \cos x$ [4]

7 Solve the equation $3 \tan \theta + 2 \sin \theta = 0$ for values of θ in the
range $0° \le \theta \le 360°$. [5]

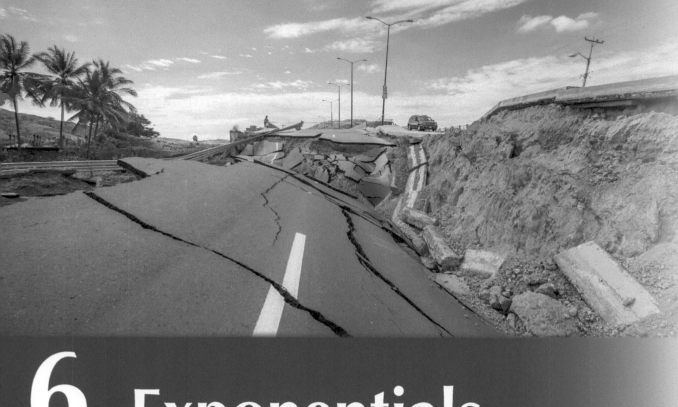

6 Exponentials and logarithms

Prior knowledge

You will need to make sure you fully understand the following from your GCSE studies:

- Sketching the graph of $y = a^x$
- The laws of indices

Quick revision

The graphs of exponential and logarithmic functions

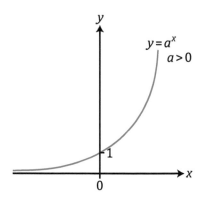

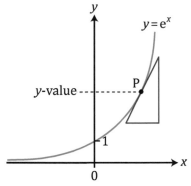

Gradient at P = y-value at P

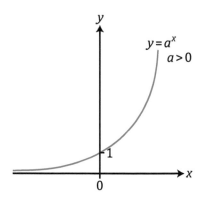

> Remember the shapes of these curves as you may have to apply transformations to them.

Logarithm – the logarithm of a positive number to a base a is the power to which the base must be raised in order to give the positive number.

If $y = a^x$ then $\log_a y = x$

> You must remember both of these equations and be able to use them.

Note that these two functions are inverses of each other so one function can be used to 'undo' the other function.

$$\log_a a = 1, \text{ as } a^1 = a$$
$$\log_a 1 = 0, \text{ as } a^0 = 1$$

The laws of logarithms

> You must be able to prove all of these laws of logarithms.

$$\log_a x + \log_a y = \log_a (xy)$$
$$\log_a x - \log_a y = \log_a \left(\frac{x}{y}\right)$$
$$k \log_a x = \log_a (x^k)$$

Solving equations in the form $a^x = b$

Equations in the form $a^x = b$ can be solved by first taking logs to base a of both sides like this:

$$a^x = b$$

Taking logarithms to base a of both sides, we obtain:

$$x = \log_a b$$

$\log_a a^x = x \log_a a = x$
As $\log_a a = 1$

Using exponential growth and decay in modelling

Exponential growth: $\quad y = ae^{kt}, \ k > 0.$

Exponential decay: $\quad y = ae^{-kt}, \ k > 0.$

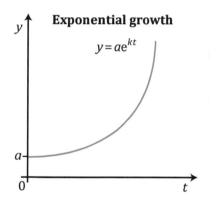

Examples of this type of growth include population growth, bacterial growth, etc.

Both of these graphs cut the y-axis at $y = a$ which is the y-value at $t = 0$ (i.e. the initial value of the quantity being modelled).

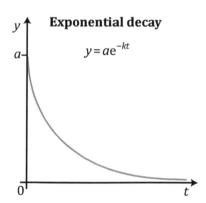

Examples of this type of decay include radioactive decay.

Using logarithmic graphs

Exponential relationships include $y = ax^n$ and $y = kb^x$ where a, b and k are parameters whose values often have to be found.

$y = ax^n$

For $y = ax^n$ and taking logs of both sides we obtain $\log y = \log a + \log x^n$ and then $\log y = \log a + n \log x$. Rearranging in the form $y = mx + c$ gives $\log y = n \log x + \log a$.

If a graph is drawn of this equation with $\log y$ on the y-axis and $\log x$ on the x-axis then the gradient $= n$ and the intercept on the y-axis is $\log a$.

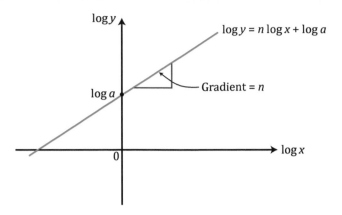

$y = kb^x$

In a similar way $\log y = x \log b + \log k$

If a graph is drawn of this equation with $\log y$ on the y-axis and x on the x-axis then the gradient $= \log b$ and the intercept on the y-axis is $\log k$.

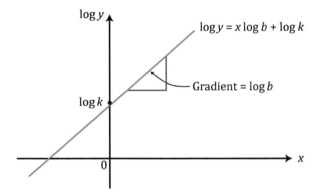

Looking at exam questions

1 (a) Solve $2 \log_{10} x = 1 + \log_{10} 5 - \log_{10} 2$. [4]

    (b) Solve $3 = 2e^{0.5x}$. [2]

    (c) Express $4^x - 10 \times 2^x$ in terms of y, where $y = 2^x$.

        Hence solve the equation $4^x - 10 \times 2^x = -16$. [5]

Thinking about the question

The rules of logs need to be used in part (a).

In part (b) notice the x is in the exponential so you will need to take natural logs of both sides.

Part (c) of the question can be solved by forming a quadratic equation.

Starting the solution

For part (a) we need to express all the numbers as logs to base 10 and then combine them and then take anti-logs of both sides. Note that the number 1 can be expressed as a log to base 10 (i.e. $1 = \log_{10} 10$).

For part (b), first rearrange to leave the exponential on its own on one of the sides. Then ln of both sides are taken and the resulting equation solved for x.

For part (c), we let $y = 2^x$ and then replace all the occurrences of 2^x in the equation by y. Note that we need to express 4^x in terms of 2^x. Once the equation in y has been solved we can then use $y = 2^x$ to find the value or values of x.

The solution

(a) $2 \log_{10} x = 1 + \log_{10} 5 - \log_{10} 2$

$\log_{10} x^2 = \log_{10} 10 + \log_{10} 5 - \log_{10} 2$

$\log_{10} x^2 = \log_{10} \left(\frac{10 \times 5}{2}\right)$

Taking antilogs of both sides, we obtain:

$x^2 = \left(\frac{10 \times 5}{2}\right)$

$x^2 = 25$

$x = 5$

> Note that $1 = \log_{10}10$

> **Watch out**
> The minus 5 solution is ignored as you cannot find the log of a negative number.

(b) $3 = 2e^{0.5x}$

$e^{0.5x} = 1.5$

Taking ln of both sides, we obtain:

$0.5x = \ln 1.5$

$x = 2 \ln 1.5$

$= 0.81093$

> In order to find the inverse of the exponential you must make sure that the exponential is on its own without a number in front. Hence here divide both sides by 0.5.

(c) $4^x - 10 \times 2^x = (2^2)^x - 10 \times 2^x$

$= 2^{2x} - 10 \times 2^x$

Replacing $2x$ with y, we obtain

$2^{2x} - 10 \times 2^x = y^2 - 10y$

Now $4^x - 10 \times 2^x = -16$

So $y^2 - 10y = -16$

$y^2 - 10y + 16 = 0$

$(y - 8)(y - 2) = 0$

$y = 8 \text{ or } 2$

> Note that $(2^2)^x = (2^x)^2$.

Note $2^3 = 8$ and $2^1 = 2$.

Now as $y = 2^x$

$$2^x = 8 \text{ or } 2$$

Hence $x = 3$ or 1

---

2   The value of a car, £V, may be modelled as a continuous variable. At time t years, the value of the car is given by $V = Ae^{kt}$, where A and k are constants. When the car is new, it is worth £30 000. When the car is two years old, it is worth £20 000. Determine the value of the car when it is six years old, giving your answer correct to the nearest £100.   [6]

Thinking about the question

The car goes down in price with time, so this question is connected with exponential decay. As there is no minus sign in the exponent (i.e. the power of e), we can conclude that the value of k must be negative. This type of question usually involves finding the values of the constants A and k and then substituting the values back into the equation. The formula containing these values can then be used.

Starting the solution

Use the pair of values to find A and k and then substitute both values back into the formula to find V when $t = 6$.

The solution

Remember $e^0 = 1$.

$$V = Ae^{kt}$$

When $V = 30\,000$, $t = 0$

Hence, $30\,000 = Ae^0$

$$A = 30\,000$$

When $V = 20\,000$, $t = 2$ so

$$20\,000 = 30\,000\,e^{2k}$$

$$e^{2k} = \frac{2}{3}$$

Taking ln of both sides, we obtain:

$$\ln\left(e^{2k}\right) = \ln\left(\frac{2}{3}\right)$$

$$2k = \ln\left(\frac{2}{3}\right)$$

$$k = \frac{1}{2}\ln\left(\frac{2}{3}\right)$$

Substituting k back into the equation, we obtain:

$$V = 30\,000\,e^{\left(\frac{1}{2}\ln\left(\frac{2}{3}\right)\right)t}$$

Now, $t = 6$, hence $V = 30\,000\,e^{\left(\frac{1}{2}\ln\left(\frac{2}{3}\right)\right)6}$

$$= \text{£8900 (nearest £100)}$$

Watch out

Always check back with the question to make sure you are giving your answer to the required accuracy.

3 The population of a new housing estate, P, can be modelled using the equation: $P = ab^n$ where a and b are constants and n is the number of years since the estate was built.

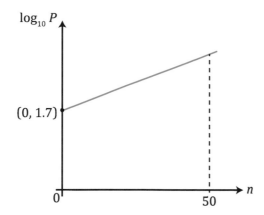

The straight line in the diagram shows how $\log_{10} P$ varies with n up to a period of 50 years.

The gradient of the line is 0.2 .

(a) Write down the equation for the line. [3]

(b) Find the value of the constants a and b. [3]

(c) State in the context of the question what each of the following means:

    (i) the intercept on the vertical axis, [1]

    (ii) the gradient of the line. [1]

(d) Find the population of the estate after a period of 10 years. [2]

(e) State two reasons why the model may not be an appropriate model to use. [2]

Thinking about the question

This question is about using logarithmic graphs to find the value of parameters. The first thing we do is to change the given equation into the form $y = mx + c$ by using the laws of logarithms. We can then use this equation to answer the rest of the question.

Starting the solution

For part (a) we take logarithms of both sides of the given equation and then use the laws of logarithms to obtain a straight-line equation. Then for part (b) we can compare the equation with the equation $y = mx + c$ to determine the gradient and the intercept. It should be noted that these will be in logarithmic form so we will use the laws of logarithms to find the values of a and b.

For (c) the intercept will be the population when $t = 0$ and the gradient will be the increase in population per year.

For (d) we can simply substitute our values for the constants a and b into the original equation and then substitute $n = 10$ to find the value of P.

For (e) we need to think about the real-life situation. For example, the estate could not keep on expanding forever.

The solution

(a) $P = ab^n$

Taking $\log_{10}$ of both sides, we obtain:

$$\log_{10} P = \log_{10} a + \log_{10} b^n$$
$$\log_{10} P = n \log_{10} b + \log_{10} a$$

> We now rearrange this equation so that it is in the form $y = mx + c$. Remember that n is a variable in this equation so $\log_{10} b$ is equivalent to m, the gradient of the line.

(b)   Gradient $= \log_{10} b$

$$\log_{10} b = 0.2$$
$$b = 10^{0.2} = 1.5849$$

Intercept $= 1.7$

Hence, $\log_{10} a = 1.7$

$$a = 10^{1.7}$$
$$= 50.119$$

(c) (i)   Intercept on the y–axis is the initial population when the estate was built (i.e. when $n = 0$).

(ii)   The gradient represents the increase in population of the estate per year.

(d) $P = ab^n$

$= 50.119 \times 1.5849^{10}$

$= 5012$

(e)   This model predicts unlimited growth.

There may be external factors such as availability of land or employment which may mean the equation can no longer be used.

Exam practice

1 (a) Given that $x > 0$, $y > 0$, show that $\log_a (xy) = \log_a x + \log_a y$. [3]

   (b) Express $\log_a 36 + \frac{1}{2} \log_a 256 - 2 \log_a 48$ as a single logarithm. [4]

   (c) Solve the equation: $2^{x+1} = 5$, giving your answer correct to three decimal places. [2]

2 (a) Given that $x > 10$, show that: $\log_a x^n = n \log_a x$. [3]

   (b) Solve the equation: $\log_a (3x + 4) - \log_a x = 3 \log_a 2$. [4]

   (c) Solve the equation: $4^{3y+2} = 7$, giving your answer correct to three decimal places. [3]

3 Solve the equation $9^x - 6(3^x) + 8 = 0$, giving the values of x correct to 3 significant figures. [5]

4 Given that $x > 0$, $y > 0$, show that $\log_a \left(\dfrac{x}{y}\right) = \log_a x - \log_a y$. [5]

5 (a) Express $\log_{10} 2 + 2 \log_{10} 18 - \frac{3}{2} \log_{10} 36$ as a single logarithm in its simplest form. [3]

   (b) Solve the equation $\log_{10} (x^2 + 48) = \log_{10} x + 2 \log_{10} 4$ [4]

6 The value of a house, £V, t years after it was built is modelled by the following equation:

$$V = ab^t \text{ where } a \text{ and } b \text{ are constants.}$$

   (a) Show that the above equation can also be written as follows:

$$\log_{10} V = t \log_{10} b + \log_{10} a$$ [2]

   (b) A graph of $\log_{10} V$ against t was plotted and a straight line l was produced.
   The gradient of l was 0.002 and the intercept on the vertical axis was 5.6.
   Use this information to find the values of a and b.
   Give each value correct to four decimal places. [4]

   (c) Explain the significance of the values of the gradient and the intercept on the vertical axis in this model. [2]

   (d) Find the value of the house as predicted by the model after 3 years from when it was first built. Give your answer to the nearest pound. [3]

   (e) It has been said the model can be used to predict the price of the house in 50 years' time. Explain why the model might not give an accurate price. [1]

7 Differentiation

Prior knowledge

You will need to make sure you fully understand the following from your GCSE studies:

- The laws of indices
- Drawing and interpreting graphs

Quick revision

Differentiating

To differentiate terms of a polynomial expression:

- Multiply by the index and then reduce the index by one.
- If $y = kx^n$ then the derivative $\dfrac{dy}{dx} = nkx^{n-1}$.

Increasing or decreasing functions

To find whether a curve or function is increasing or decreasing at a certain point:

- Differentiate the equation of the curve or the function.

- Substitute the value of x at the given point into the expression for the derivative to see whether the gradient is positive or negative. If the value is positive, the function is increasing at the given point, and if the value is negative, the function is decreasing.

Finding a stationary point

- Substitute $\frac{dy}{dx} = 0$ and solve the resulting equation to find the value or values of x at stationary points.

- Substitute the value or values of x into the equation of the curve to find the corresponding y-coordinate(s).

Finding whether a stationary point is a maximum or minimum

- Differentiate the first derivative $\left(\text{i.e. } \frac{dy}{dx}\right)$ to find the second derivative $\left(\text{i.e. } \frac{d^2y}{dx^2}\right)$.

- Substitute the x-coordinate of the stationary point into the expression for $\frac{d^2y}{dx^2}$.

- If the resulting value is negative then the stationary point is a maximum point and if the resulting value is positive, then the stationary point is a minimum point. If $\frac{d^2y}{dx^2} = 0$, then the result is inconclusive and further investigation is required.

Determining whether a stationary point is a point of inflection

- Substitute the x-coordinate of a point either side of the stationary point into the expression for $\frac{dy}{dx}$ and if the gradient has the same sign then the stationary point is a point of inflection.

Curve sketching

- Find the points of intersection with the x and y axes by substituting $y = 0$ and $x = 0$ in turn and then solving the resulting equations.

- Find the stationary points and their nature (i.e. maximum, minimum, point of inflection).

- Plot the above on a set of axes.

- It is important to note that maxima and minima are local maximum and minimum points only, and not necessarily the maximum or minimum values for a function. For example, the graph of a cubic equation shows this clearly.

Looking at exam questions

1   Given that $y = x^3$, find $\dfrac{dy}{dx}$ from first principles.   [6]

Thinking about the question

This question requires you to remember the steps involved in differentiating from first principles.

Starting the solution

We need to increase the x and y values by small amounts δx and δy respectively and then put them into the equation. We must make sure that we mention the limit when δx approaches zero.

The solution

Increasing x by a small amount δx will result in y increasing by a small amount δy.

Putting $x + \delta x$ and $y + \delta y$ into the equation we obtain:

$$y + \delta y = (x + \delta x)^3$$
$$= (x + \delta x)(x + \delta x)(x + \delta x)$$
$$= (x + \delta x)(x^2 + 2x\delta x + \delta x^2)$$
$$= x^3 + 3x^2\delta x + 3x(\delta x)^2 + (\delta x)^3$$

Now $y = x^3$, so substituting this for y on the left-hand side we obtain:

$$x^3 + \delta y = x^3 + 3x^2\delta x + 3x\delta x^2 + (\delta x)^3$$
$$\delta y = 3x^2\delta x + 3x(\delta x)^2 + (\delta x)^3$$

Dividing both sides by δx,

$$\frac{\delta y}{\delta x} = 3x^2 + 3x\delta x + (\delta x)^2$$

Letting $\delta x \to 0$

$$\frac{dy}{dx} = \lim_{\delta x \to 0} \frac{\delta y}{\delta x} = 3x^2$$

Watch out

Not including the part about the limit will cost you a mark in the exam.

---

2   (a)   Given that $y = \dfrac{5}{x} + 6\sqrt[3]{x}$, find $\dfrac{dy}{dx}$ when $x = 8$.   [3]

   (b)   Find the range of values for which $f(x) = \dfrac{2}{3}x^3 + \dfrac{5}{2}x^2 - 3x + 1$ is an increasing function.   [4]

Thinking about the question

Part (a) is a simple differentiation question.

Part (b) talks about an increasing function and this means that the y-value keeps on increasing as the x-value increases.

Starting the solution

For part (a) we need to write the equation of the curve in terms of indices and then differentiate and substitute $x = 8$ into the resulting $\dfrac{dy}{dx}$.

For part (b) we are looking for the range of values of x where the gradient is positive. We therefore differentiate the function and then create an inequality by making the differential greater than zero. We can then solve the resulting inequality to find the range of values of x.

The solution

(a) $y = \dfrac{5}{x} + 6\sqrt[3]{x}$

$\qquad = 5x^{-1} + 6x^{\frac{1}{3}}$

$$\dfrac{dy}{dx} = -5x^{-2} + 2x^{-\frac{2}{3}}$$

$$= -\dfrac{5}{x^2} + \dfrac{2}{\sqrt[3]{x^2}}$$

When $x = 8$, $\dfrac{dy}{dx} = -\dfrac{5}{8^2} + \dfrac{2}{\sqrt[3]{8^2}}$

$$= -\dfrac{5}{64} + \dfrac{1}{2}$$

$$= \dfrac{27}{64}$$

(b) $f(x) = \dfrac{2}{3}x^3 + \dfrac{5}{2}x^2 - 3x + 1$

$\qquad\qquad\qquad\qquad\qquad f'(x) = 2x^2 + 5x - 3$

For an increasing function $\qquad f'(x) > 0$

Hence, $\qquad\qquad\qquad\qquad 2x^2 + 5x - 3 > 0$

Finding the critical values $\qquad 2x^2 + 5x - 3 = 0$

$\qquad\qquad\qquad\qquad\qquad (2x - 1)(x + 3) = 0$

So $x = \dfrac{1}{2}$ or $x = -3$

> Remember that as the number in front of the x^2 term is positive, the curve will be U-shaped intersecting the x-axis at the roots $x = \dfrac{1}{2}$ and $x = -3$.

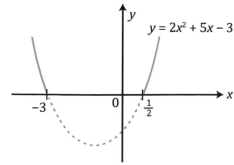

$y = 2x^2 + 5x - 3$

> As $2x^2 + 5x - 3 > 0$ we want the values of x where the curve is above the x-axis (shown as bold on the graph).

Range is $x < -3$ or $x > \dfrac{1}{2}$

3   The curve C has equation $y = 7 + 13x - 2x^2$. The point P lies on C and is such that the tangent to C at P has equation $y = x + c$, where c is a constant. Find the coordinates of P and the value of c.   [5]

Thinking about the question

This question is about the tangent to a curve. We can find the gradient of the curve at any point by differentiating the equation of the curve.

Starting the solution

As the line $y = x + c$ is a tangent to the curve, we can compare it with the equation of a straight line $y = mx + c$ and we see that the gradient, m, is 1.

Once we have differentiated the equation of the curve, we can put the result equal to 1 and then solve the equation to find x. We can then put this x-coordinate into the equation of the curve to find the y-coordinate.

We can then put the coordinates of P back into the equation of the tangent $y = x + c$ to find the value of c.

The solution

$$y = 7 + 13x - 2x^2$$

$$\frac{dy}{dx} = 13 - 4x$$

> Remember to substitute the x-coordinate back into the equation to obtain the y-coordinate.

Equation of the tangent at P is $y = x + c$ so the gradient of the tangent = 1.

Hence $13 - 4x = 1$ giving $x = 3$.

When $x = 3$, $y = 7 + 13(3) - 2(3)^2 = 28$

P is the point (3, 28)

> Since P lies on the line, its coordinates will satisfy the equation and this can be used to find the value of c.

Equation of tangent at P is $y = x + c$ so when $x = 3$ and $y = 28$ we have:

$$28 = 3 + c \text{ giving } c = 25.$$

4   A curve C has equation $y = x^3 - 3x^2$.

  (a) Find the stationary points of C and determine their nature.   [7]

  (b) Draw a sketch of C, clearly indicating the stationary points and the points where the curve crosses the coordinate axes.   [3]

  (c) **Without performing the integration**, state whether

$$\int_0^3 (x^3 - 3x^2)dx$$

  is positive or negative, giving a reason for your answer.   [1]

Thinking about the question

The first two parts of this question involve finding the stationary points and then using this information along with the points of intersection of the curve with both axes, in order to sketch the graph.

For part (c) we have to use the graph to work out whether there is more of the graph above or below the x-axis between the two limits. It is hard to see how this will work, without drawing the graph.

Starting the solution

Part (a) involves differentiating and then equating this to zero and solving the resulting equation to determine the x-coordinates of the stationary points. We then need to substitute these back into the equation of the curve to find the corresponding y-coordinates.

To determine whether each point is a max or min, we differentiate again and then substitute in the x-coordinates in turn to see the sign of the result. If it is negative the point is a max and if positive, then the point is a min.

Part (b) involves substituting $x = 0$ into the equation to find the point of intersection with the y-axis. Then we substitute $y = 0$ and solve the equation to find the point/points of intersection with the x-axis. We can use this along with the information about stationary points to draw a sketch of the curve.

Part (c) involves looking at the curve to see if some or all of it lies below the x-axis. Remember that those parts lying below the axis will have a negative area.

The solution

(a) $y = x^3 - 3x^2$

$$\frac{dy}{dx} = 3x^2 - 6x$$

At the stationary points, $\quad \dfrac{dy}{dx} = 0$

Hence, $\qquad\qquad 3x^2 - 6x = 0$

$$3x(x - 2) = 0$$

$x = 0$ or 2

when $x = 0$, $\qquad\qquad y = (0)^3 - 3(0)^2 = 0$

when $x = 2$, $\qquad\qquad y = (2)^3 - 3(2)^2 = -4$

$$\frac{d^2x}{dy^2} = 6x - 6$$

Hence when $x = 0$, $\qquad \dfrac{d^2x}{dy^2} = 6(0) - 6 = -6 < 0$

> The x-coordinates are substituted into the equation of the curve to find the corresponding y-coordinates.

67

showing there is a maximum point at $x = 0$.

When $x = 2$, $\qquad\qquad \dfrac{d^2y}{dx^2} = 6(2) - 6 = 6 > 0$

showing there is a minimum point at $x = 2$.

Hence $(0, 0)$ is a maximum point and $(2, -4)$ is a minimum point.

(b) When $y = 0$, $x^3 - 3x^2 = 0$ so $x^2(x - 3) = 0$, giving $x = 0$ or 3.

When $x = 0$, $y = 0$ (meaning the curve passes through the origin).

> Remember to label all the points.

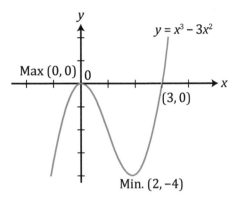

(c) The section of the curve between $x = 0$ and $x = 3$ lies below the x-axis and so the area is negative in this region and the integral for this interval would be negative.

---

5 Curve C has equation $y = 2x^2 + 6x + 7$. Point P whose x-coordinate is -1 lies on C. The tangent to the curve at P has the equation $y = mx + c$. Find the value of m and the value of c. [5]

Thinking about the question

This question concerns the tangent to a curve. The expression for the gradient of the curve can be found using differentiation.

Starting the solution

We can find the expression for the gradient of the curve by differentiating its equation. The value at $x = -1$ can then be found which will also be the gradient of the tangent. We can then find the y-coordinate by substituting the x-coordinate into the equation of the curve $y = 2x^2 + 6x + 7$.

We can finally substitute the coordinates for P and the gradient m into the equation of the line $y = mx + c$ to find the value of c.

The solution

$$y = 2x^2 + 6x + 7$$

$$\frac{dy}{dx} = 4x + 6$$

At $x = -1$, $\quad \frac{dy}{dx} = 4(-1) + 6 = 2$

To find the y-coordinate of P: $y = 2(-1)^2 + 6(-1) + 7 = 3$

Hence P has coordinates $(-1, 3)$

The tangent has gradient of 2 and passes through $(-1, 3)$

Now $\qquad\qquad\qquad y = mx + c$

so $\qquad\qquad\qquad 3 = 2(-1) + c$

$$c = 5$$

Hence $m = 2$ and $c = 5$.

Exam practice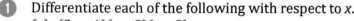

① Differentiate each of the following with respect to x.
(a) $(2x - 1)(x + 3)(x + 2)$ [2]
(b) $\sqrt{x}\left(\sqrt{x} - x^2\right)$ [2]

② If $y = 6\sqrt{x} + \dfrac{25}{x^2} + 3x^5$, find the gradient of the curve where $x = 1$. [3]

③ (a) Given that $y = (2x - 1)^2$ find $\dfrac{dy}{dx}$ from first principles. [6]

(b) Find the range of values of x for which the function,
$f(x) = (2x - 1)^2$ is an increasing function. [1]

④ Differentiate $\dfrac{3}{x^4} + 4\sqrt{x}$ with respect to x. [4]

⑤ Curve C has equation $y = x^3 - 3x^2 - 9x + 7$.
(a) Find the stationary points of C and determine their nature. [7]

(b) Draw a sketch of C, clearly indicating the stationary points
and the point where the curve crosses the y-axis. [3]

⑥ The curve C has equation $y = 3x^{\frac{3}{2}} - \dfrac{32}{x}$.
Find the equation of the normal to C at the point where $x = 4$. [7]

7   A rectangular sheet of metal has length 8 m and width 5 m. Four squares, each of side x m, where $x < 2.5$, have been cut away from the corners of the rectangular sheet, as shown in the diagram below. The rest of the metal sheet is now bent along the dotted lines to form an open tank in the form of a cuboid.

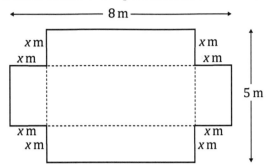

(a) Show that the volume V m³ of this tank is given by
$$V = 4x^3 - 26x^2 + 40x$$ [2]

(b) Find the maximum value of V, showing that the value you have found is a maximum value. [5]

8   Differentiate each of the following with respect to x.

(a) $2x^5 + \dfrac{24}{x^2} - \sqrt[3]{x}$ [2]

(b) $\sqrt{x}\left(\sqrt[3]{x} - 2x + 1\right)$ [3]

9   The curve C has equation: $y = x^4 + x + 1$
Find the equation of the tangent to C at the point $(1, 3)$. [4]

10   The curve C has equation
$$y = x^3 - 3x^2 - 9x + 3.$$
(a) Find the coordinates and nature of the stationary points of C. [8]
(b) Sketch C. [3]

8 Integration

Prior knowledge

You will need to make sure you fully understand the following from your GCSE studies:

- Indices
- Sketching curves

Quick revision

Indefinite integration is the reverse process to differentiation. When integrating indefinitely you must remember to include the constant of integration.

$$\int x^n \, dx = \frac{x^{n+1}}{n+1} + c \qquad \text{(provided } n \neq -1\text{)}$$

Definite integration is integration where you have limits. The answer to a definite integral will usually be a numerical value.

A definite integral is positive for areas above the x-axis and negative for areas below the x-axis.

A final area must always be given as a positive value.

Looking at exam questions

1 Integrate $5x^{\frac{1}{3}} + 3x^{-3}$ with respect to x. [2]

Thinking about the question

This is a simple integration with no limits, so a constant of integration needs to be included.

Starting the solution

We need to integrate each term by increasing the index by 1 and then dividing by the new index and then include a constant of integration, c.

The solution

$$\int \left(5x^{\frac{1}{3}} + 3x^{-3}\right)dx = \frac{5x^{\frac{4}{3}}}{\frac{4}{3}} + \frac{3x^{-2}}{-2} + c$$

$$= \frac{15}{4}x^{\frac{4}{3}} - \frac{3}{2}x^{-2} + c$$

$$= \frac{15}{4}x^{\frac{4}{3}} - \frac{3}{2x^2} + c$$

> **Watch out**
>
> It is common to see students getting mixed up between integration and differentiation.

> Always remember to include a constant of integration when there are no limits.

---

2 Integrate $\sqrt{x} - \frac{2}{x^2}$ with respect to x. [4]

Thinking about the question

This is indefinite integration so we need to include a constant. We also need to use index form for the occurrences of x before we integrate.

Starting the solution

$\sqrt{x}$ can be expressed as $x^{\frac{1}{2}}$ and $\frac{2}{x^2}$ can be expressed as $2x^{-2}$.

We need to be particularly careful of the signs during the integration. When you increase -2 by 1 it becomes -1.

The solution

$$\int \left(\sqrt{x} - \frac{2}{x^2}\right)dx = \int \left(x^{\frac{1}{2}} - 2x^{-2}\right)dx$$

$$= \frac{x^{\frac{3}{2}}}{\frac{3}{2}} - \frac{2x^{-1}}{-1} + c$$

$$= \frac{2}{3}x^{\frac{3}{2}} + 2x^{-1} + c$$

$$= \frac{2}{3}\sqrt{x^3} + \frac{2}{x} + c$$

> Remember when you have a fraction as a denominator you turn the fraction upside down and multiply the numerator by it.

> For all indefinite integration (i.e. when there are no limits) we need to include a constant of integration, c.

3

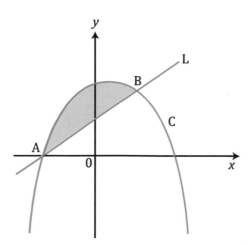

The sketch shows the curve C with equation $y = 14 + 5x - x^2$ and line L with equation $y = x + 2$. The line intersects the curve at the points A and B.

(a)   Find the coordinates of A and B.   [4]

(b)   Calculate the area enclosed by L and C.   [6]

Thinking about the question

1   For part (a) you simply solve the equation of the line with the equation of the curve to find the coordinates of the points of intersection.

2   For part (b) it will be necessary to find the area under the curve between A and B using integration and then find the area of the triangle which is subtracted from the area under the curve to find the shaded area.

Starting the solution

For part (a) we need to find the coordinates of A and B. To do this, we equate the equation of the line with that of the curve and solve the resulting equation. This gives us the x-coordinates of points A and B which can be substituted into either equation to find the corresponding y-coordinates.

For part (b) we need to integrate the equation of the curve and use the x-coordinates of points A and B as the limits. We then find the area of the triangle formed by AB and the x-axis and the perpendicular from the x-axis to B.

The area of the triangle is subtracted from the area under the curve to given the required area which is shown shaded in the diagram.

At the points of intersection both equations will have the same y-values.

Remember to state which of the coordinates is A and which is B.

The solution

(a) $14 + 5x - x^2 = x + 2$

$$x^2 - 4x - 12 = 0$$

$$(x - 6)(x + 2) = 0$$

Hence $x = 6$ or -2

$y = x + 2$, so when $x = 6$, $y = 6 + 2 = 8$
and when $x = -2$, $y = -2 + 2 = 0$

Hence A is $(-2, 0)$ and B is $(6, 8)$

(b) Area under the curve between A and B

$$= \int_{-2}^{6} y \, dx$$

$$= \int_{-2}^{6} \left(14 + 5x - x^2\right) dx$$

$$= \left[14x + \frac{5x^2}{2} - \frac{x^3}{3}\right]_{-2}^{6}$$

$$= \left[\left(84 + 90 - 72\right) - \left(-28 + 10 + \tfrac{8}{3}\right)\right]$$

$$= \frac{352}{3}$$

Area of triangle $= 0.5 \times 8 \times 8 = 32$

Shaded area $= \dfrac{352}{3} - 32$

$$= 85\tfrac{1}{3} \text{ sq units}$$

---

4 The diagram shows a sketch of the curve $y = 2 + x - x^2$.

The curve intersects the x-axis at B and C. Point A has coordinates $(-2, 0)$.

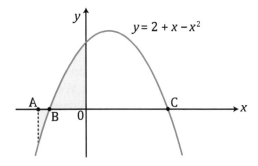

Find the **total area** of the shaded region.

Thinking about the question

There are two areas shaded; one above the x-axis and one below the x-axis.

To find areas it is necessary to integrate the equation of the curve and put in limits.

Starting the solution

The area above the x-axis will be positive and that below the x-axis will be negative. It is therefore necessary to find these two areas separately and change the negative area to a positive area before adding the two areas together.

In order to find the areas, we need to find the x-coordinate of point B and this can be found by equating the equation of the curve to zero and then solve the quadratic equation. There will be two solutions, one for B and the other for C.

We can then do two integrations using the limits and disregard the negative sign for the area below the x-axis and then add the two areas together to give the total area.

The solution

$$y = 2 + x - x^2$$

Where the curve cuts the x-axis, $y = 0$ so

$$0 = 2 + x - x^2$$

$$x^2 - x - 2 = 0$$

$$(x - 2)(x + 1) = 0$$

Hence, $x = -1$ or 2

This means B is $(-1, 0)$

Area between A and B $= \int_{-2}^{-1} \left(2 + x - x^2\right) dx$

$$= \left[2x + \frac{x^2}{2} - \frac{x^3}{3}\right]_{-2}^{-1}$$

$$= \left[\left(-2 + \frac{1}{2} + \frac{1}{3}\right) - \left(-4 + 2 + \frac{8}{3}\right)\right]$$

$$= -\frac{11}{6}$$

Area between B and O $= \int_{-1}^{0} \left(2 + x - x^2\right) dx$

$$= \left[2x + \frac{x^2}{2} - \frac{x^3}{3}\right]_{-1}^{0}$$

It is easier to factorise this quadratic equation if you swap the sides to make x^2 positive.

75

$$= \left[\left(0 + 0 - 0 \right) - \left(-2 + \frac{1}{2} + \frac{1}{3} \right) \right]$$

$$= \frac{7}{6}$$

Shaded area $= \dfrac{11}{6} + \dfrac{7}{6} = \dfrac{18}{6} = 3$

Exam practice

1 Find $\displaystyle\int \left(4x^3 - \frac{4}{x^2} + 6\sqrt{x} - 1 \right) dx$ [3]

2 The gradient of curve C is given by: $\dfrac{dy}{dx} = 3x^2 + 6x - 1$

   Given that curve C passes through the point $(1, 7)$, find the
   equation of the curve.   [5]

3 The gradient of a curve is given by $\dfrac{dy}{dx} = x^2 + 2x - 8$.

   The curve passes through the point $P(3, 0)$.

   (a) Show that the equation of the curve is $y = \dfrac{x^3}{3} + x^2 - 8x + 6$.   [3]

   (b) Find the coordinates of the stationary points of this curve.   [3]
   (c) Sketch the curve showing the stationary points and the
        intercept on the y-axis.   [4]

4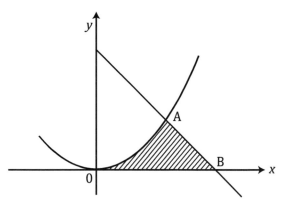

   The diagram shows the curve $y = 2x^2$ and the line $y = 12 - 2x$
   intersecting at the point A. The line $y = 12 - 2x$ intersects the
   x-axis at B.
   (a) Find the coordinates of A and B.   [5]
   (b) Evaluate the area of the shaded region.   [7]

5 (a) Find $\displaystyle\int\left(2x^{\frac{3}{4}}+\frac{7}{x^{\frac{1}{2}}}\right)dx.$ [2]

(b)

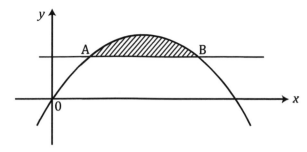

The diagram shows a sketch of the curve $y = 6x - x^2$ and the line $y = 5$. The line and the curve intersect at the points A and B. Find the area of the shaded region. [10]

6 Integrate $\sqrt[3]{x^2}\,(x-1)$ with respect to x. [3]

9 Vectors

Prior knowledge

You will need to make sure you fully understand the following from your GCSE studies:

- Ratios
- Properties of quadrilaterals

Quick revision

Scalar – a quantity that has size only (e.g. distance, speed and time).

Vector – a quantity that has both size and direction (e.g. displacement, velocity and force).

Condition for two vectors to be parallel

For two vectors **a** and **b** to be parallel

$$\mathbf{a} = k\mathbf{b}$$

where k is a scalar

The magnitude of a vector

The vector $\mathbf{r} = a\mathbf{i} + b\mathbf{j}$ has magnitude given by $|\mathbf{r}| = \sqrt{a^2 + b^2}$

The distance between two points

The distance between two points A (x_1, y_1) and B (x_2, y_2) is given by:

$$d = \sqrt{(x_2 - x_1)^2 + (y_2 - y_1)^2}$$

The position vector of a point dividing a line in a given ratio

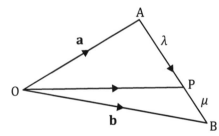

Point P dividing AB in the ratio $\lambda : \mu$ has position vector $\overrightarrow{OP}$

where
$$\overrightarrow{OP} = \frac{\mu\mathbf{a} + \lambda\mathbf{b}}{\lambda + \mu}$$

Looking at exam questions

1   Points P (6, 3), Q (2, 7) and S (1, 2) are the coordinates of three of the corners of the rhombus PSQT.

(a)   Find the position vector of M, the mid-point of PQ.   [2]

(b)   Find the position vector of T.   [3]

Thinking about the question

This question is about vectors. As always, it is important to mark the coordinates on a graph.

Starting the solution

Part (a) is a coordinate geometry question that involves the use of the mid-point formula, which needs to be recalled from memory.

For part (b) we need to know that the opposite sides of a rhombus are parallel and the same length. The vectors (not the position vectors) of the opposite sides will be the same.

The solution

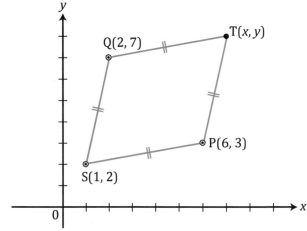

(a) Mid-point of PQ $= \left(\dfrac{x_1 + x_2}{2}, \dfrac{y_1 + y_2}{2}\right) = \left(\dfrac{2 + 6}{2}, \dfrac{7 + 3}{2}\right) = (4, 5)$

(b) $\overrightarrow{SQ} = \mathbf{i} + 5\mathbf{j}$

As $\overrightarrow{SQ}$ and $\overrightarrow{PT}$ are parallel and the same length they will have the same vector, so

$$\overrightarrow{PT} = \mathbf{i} + 5\mathbf{j}$$
$$\overrightarrow{PT} = \overrightarrow{PO} + \overrightarrow{OT}$$
$$\mathbf{i} + 5\mathbf{j} = -6\mathbf{i} - 3\mathbf{j} + \overrightarrow{OT}$$
$$\overrightarrow{OT} = 7\mathbf{i} - 8\mathbf{j}$$

Position vector of T $= 7\mathbf{i} - 8\mathbf{j}$

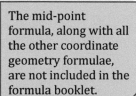

The mid-point formula, along with all the other coordinate geometry formulae, are not included in the formula booklet.

Note that position vectors of a point are the vectors from O to the point. Hence $\overrightarrow{OT}$ is the position vector of point T.

---

2 (a) The vectors $\mathbf{u}$ and $\mathbf{v}$ are defined by $\mathbf{u} = 9\mathbf{i} - 40\mathbf{j}$ and $\mathbf{v} = 3\mathbf{i} - 4\mathbf{j}$. Determine the range of values for μ such that $\mu|\mathbf{v}| > |\mathbf{u}|$.   [3]

(b) The point A has position vector $11\mathbf{i} - 4\mathbf{j}$ and the point B has position vector $21\mathbf{i} + \mathbf{j}$. Determine the position vector of the point C, which lies between A and B, such that AC:CB is 2:3.   [3]

Thinking about the question

Part (a) is concerned with the magnitude of vectors and there is a formula for this that you need to remember.

Part (b) concerns the position vector of a point dividing a line in a given ratio and the formula for this can be looked up in the formula booklet.

Starting the solution

For part (a) we can find the magnitude of vectors **u** and **v** and then put them into the given equality and then solve this for find the range of values for μ.

For part (b) we look up the following formula: The point P dividing AB in the ratio $\lambda:\mu$ has position vector

$$\overrightarrow{OP} = \frac{\mu\mathbf{a} + \lambda\mathbf{b}}{\mu + \lambda}$$

and then enter the numbers for the variables.

The solution

    (a) $|\mathbf{u}| = \sqrt{9^2 + (-40)^2} = \sqrt{1681} = 41$

          $|\mathbf{v}| = \sqrt{3^2 + (-4)^2} = \sqrt{25} = 5$

          Now, $\mu|\mathbf{v}| > |\mathbf{u}|$ so $5\mu > 41$

          Hence $\mu > \dfrac{41}{5}$ so $\mu > 8.2$

    (b)

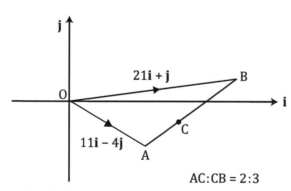

                                      AC:CB = 2:3

AB is divided in the ratio 2:3

From the formula booklet: The point P dividing AB in the ratio $\lambda:\mu$ has position vector $\overrightarrow{OP} = \dfrac{\mu\mathbf{a} + \lambda\mathbf{b}}{\lambda + \mu}$

As $\lambda = 2$ and $\mu = 3$

$$\overrightarrow{OP} = \frac{\mu\mathbf{a} + \lambda\mathbf{b}}{\mu + \lambda}$$

$$= \frac{3(11\mathbf{i} - 4\mathbf{j}) + 2(21\mathbf{i} + \mathbf{j})}{2 + 3}$$

$$= \frac{33\mathbf{i} - 12\mathbf{j} + 42\mathbf{i} + 2\mathbf{j}}{5}$$

$$= \frac{75\mathbf{i} - 10\mathbf{j}}{5}$$

$$= 15\mathbf{i} - 2\mathbf{j}$$

Exam practice

1

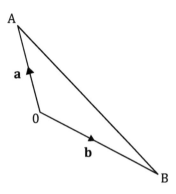

The position vectors of points A and B are:

$$\mathbf{a} = -2\mathbf{i} + 5\mathbf{j}$$
$$\mathbf{b} = 9\mathbf{i} - 2\mathbf{j}$$

(a) Find $\overrightarrow{AB}$ [1]

(b) Point P divides the line AB in the ratio 2:1.
    Find the position vector $\overrightarrow{OP}$.   [3]

2   The position vectors of points A, B and C are $\mathbf{i} + 3\mathbf{j}$, $3\mathbf{i} + 7\mathbf{j}$ and $4\mathbf{i} + 9\mathbf{j}$ respectively.

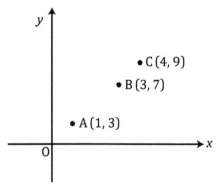

(a) (i)   Find the vectors $\overrightarrow{AB}$ and $\overrightarrow{BC}$.   [2]
     (ii) What can be deduced about the points A, B and C?   [1]

(b) Find the ratio AB:BC.   [2]

3 The point A has position vector $-12\mathbf{i} - 5\mathbf{j}$ and the point B has position vector $10\mathbf{i} + 6\mathbf{j}$. Determine the position vector of the point P, which lies between A and B, such that AP:PB is 8:3. [3]

4 OABC is a rhombus with $\overrightarrow{OA} = \mathbf{a}$ and $\overrightarrow{OC} = \mathbf{c}$. D is the mid-point of BC and the point E lies on AB such that AE:EB = 2:1.

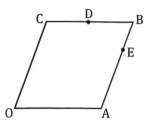

(a) (i) Express $\overrightarrow{OB}$ in terms of $\mathbf{a}$ and $\mathbf{c}$. [1]

    (ii) Express $\overrightarrow{DE}$ in terms of $\mathbf{a}$ and $\mathbf{c}$. [3]

(b) Prove that DE and CE are not parallel to each other. [2]

5 OAB is a triangle. Point A has position vector $10\mathbf{i} + 2\mathbf{j}$ and point B has position vector $8\mathbf{i} + 8\mathbf{j}$. M is the mid-point of AB and Q is the mid-point of OA.

(a) Find $\overrightarrow{QM}$. [3]

(b) Prove that QM and OB are parallel. [3]

10 Statistical sampling

Prior knowledge

You will need to make sure you fully understand the following from your GCSE studies:

- Collecting data to test a hypothesis

- Sampling methods

- Stratified samples and random samples

- Considering the effect of sample size and other factors that affect the reliability of conclusions drawn

Quick revision

Population – all members of the set that is being studied or has data collected

Sample – a smaller subset of the population and is used to draw conclusions about the population

Sampling techniques

Simple random sampling – each item in the population is given a number and then the required number of items in the sample are picked at random using a calculator, program or website.

Systematic sampling – the sampling interval is found by dividing the population by the sample size. You then pick a random number within this sample size and start from that as the first item in the sample. You then add the sampling interval to the random number to get the next number in the sample. This is repeated until you have the required sample.

Opportunity sampling – here you decide on the sample size and simply use the most convenient way of collecting the sample (e.g. friends, relatives, classmates, work colleagues, etc).

The ideal sample will:

- be large enough
- represent the population
- be unbiased.

Sampling methods advantages and disadvantages

	Simple random	Systematic	Opportunity
Advantages	Least biased of all sampling techniques	More straightforward compared to random sampling	Easy to take the sample as it is drawn from that part of the population that is close at hand
	Easily performed using website or calculator	Sample is easy to select	
	Sample is highly representative of the population		

	Simple random	Systematic	Opportunity
Disadvantages	Time consuming and tedious to perform	Not as random compared to random sampling	Sample can be highly unrepresentative of the population, as the sample is not picked at random
	Poor representation of the overall population, if certain members are not hit by the random numbers generated		

Looking at exam questions

1 Joshua collects data from the 20 students in his class at school about the number of text messages sent per day for his class and the results he obtained are shown here:

12, 50, 76, 23, 100, 65, 29, 10, 14, 30, 56, 45, 19, 21, 26, 60, 21, 18, 78, 20

(a) (i) Explain what is meant by an opportunity sample. [1]

(ii) Take an opportunity sample of the first 10 numbers in the above list and calculate the mean number of text messages per day. [2]

(b) You are given the following, using the population, for the above set of data.

$$\sum x = 773, \quad \sum x^2 = 42\,699 \quad \text{and } n = 20.$$

(i) Explain using information from the question, the difference between a sample and a population. [2]

(ii) Using these figures, find the mean number of text messages per day for his class using the population rather than a sample. [3]

(iii) Find the standard deviation for the number of text messages per day, giving your answer to one decimal place. [2]

Thinking about the question

Part (a) of the question is about an opportunity sample and as such it only uses part of the data.

Part (b) involves using data from the whole population to find the mean and standard deviation.

Starting the solution

For part (a) we need to think about what the population is. As he is only looking at his class and is not considering teenagers in general, the class is the population and any smaller part of this will be the sample.

For part (b) we are using the information only from this part of the question. This means we do not use the mean calculated from part (a) as this was based on a sample.

We need to use $\mu = \frac{\sum x}{n}$ to work out the population mean and then use

Variance $= \frac{\sum x_i^2}{n} - \left(\frac{\sum x_i}{n}\right)^2$ and then standard deviation, $\sigma = \sqrt{\text{variance}}$.

Where $\frac{\sum x_i^2}{n}$ is the mean of the squares of the values and $\left(\frac{\sum x_i}{n}\right)^2$ is the square of the mean of the values.

The solution

(a) (i) A sample drawn from the population that is close at hand and easy to take.

(ii) Mean number of text messages sent per day

$$= \frac{12 + 50 + 76 + 23 + 100 + 65 + 29 + 10 + 14 + 30}{10}$$

$$= \frac{409}{10}$$

$$= 40.9$$

(b) (i) The population is all the members of the set being studied. Here it would be all the students in the class.

The sample is a smaller subset of the population that can be used to draw conclusions about the population.

(ii) $\mu = \frac{\sum x}{n} = \frac{773}{20} = 38.65$

This formula is not included in the formula booklet and so will need to be recalled from memory.

(iii) Variance $= \frac{\sum x_i^2}{n} - \left(\frac{\sum x_i}{n}\right)^2 = \frac{42\,699}{20} - 38.65^2 = 641.1275$

This formula needs to be remembered.

Standard deviation $= \sqrt{\text{variance}} = \sqrt{641.1275}$

This formula needs to be remembered.

$$= 25.32$$

Exam practice

① There are 1500 students in a large secondary school. The head-teacher of the school would like to investigate the students' opinions of the sports facilities in the school. The head-teacher decides to use a sample of 50 students to get their opinions. Describe the steps the head-teacher should take in order to obtain a simple random sample of 50 students from the total of 1500 students. [5]

② A researcher is researching the amount of time students in a year 12 maths class at a school spend on social media sites on a Saturday.
The results in hours are shown here:
2, 4, 0, 1, 3, 2, 5, 7, 3, 4, 2, 1, 0, 8, 5, 3, 2, 1, 1, 4, 2, 1, 0, 1, 6
(a) Taking an opportunity sample of the first 5 numbers in the list, calculate the mean number of hours spent on social media sites on Saturday. [3]
(b) A systematic sample is to be taken using 5 data values.
    (i)   Work out the sampling interval. [1]
    (ii) A random number was chosen in the sampling interval and it was found to be 2. Using this value, write down the list of data values in the sample. [3]
    (iii) Using the list you produced in part (ii), work out the mean number of hours spent on social media sites on Saturday. [2]
(c) State, giving reasons, which sampling method is likely to give more reliable results. [3]

③ The total household earnings were recorded along a street with 30 houses and the following results were obtained where the figures are in pounds to the nearest thousand.
32 000, 28 000, 20 000, 63 000, 18 000, 29 000, 50 000, 18 000, 74 000, 48 000
15 000, 67 000, 23 000, 94 000, 67 000, 89 000, 65 000, 87 000, 85 000, 69 000
37 000, 55 000, 86 000, 78 000, 31 000, 42 000, 79 000, 40 000, 89 000, 46 000

(a) An opportunity sample was taken using the first 10 values in the list.
    (i)   Describe what is meant by an opportunity sample. [2]
    (ii) Calculate the mean household earnings using this sample. [3]
(b) A systematic sample is to be taken of 6 data values.
    (i)   Work out the sampling interval. [2]
    (ii) A random number was picked in the sampling interval and it was 4. Using this value write down the list of numbers in the sample. [2]

      (iii) Using the numbers in your sample, work out the mean
           household earnings. [2]

      (iv) Explain the reason why the systematic sample is giving
           a much higher mean household income compared to
           the opportunity sample. [1]

4   Jack, a statistics student, is interested to find out which is the
most popular sport viewed on TV by the students in his school.
There are 1200 pupils in his school so he decides to ask a
smaller sample of 30 students in his class and the results are
shown here:

          Football   20
          Rugby   12
          Boxing   12
          Tennis   6

(a) Using this example, explain the difference between the
    population and a sample. [2]

(b) Explain two ways in which Jack's sample may not represent
    the population. [2]

(c) Here are some informal inferences made about the
    population. For each one, say with a reason, whether the
    inference is justified or not.

    (i)   Football is more popular than boxing. [1]

    (ii) Rugby and boxing are equally popular. [1]

11 Data presentation and interpretation

Prior knowledge

You will need to make sure you fully understand the following from your GCSE studies:

- The difference between quantitative and qualitative data

- The difference between discrete and continuous data

- Constructing and interpreting scatter diagrams

- Constructing cumulative frequency diagrams

- Mean, median and mode

- Measures of spread; range and interquartile range

- Box and whisker diagrams

- Equations of straight lines of the form $y = mx + c$

Quick revision

Histograms

There are no gaps between the bars, and the height of each bar is the frequency density.

The area of the bar is equal to the frequency, where

Frequency (area of bar) = frequency density × class width

To find the height of a bar use

Frequency density (i.e. height) = $\dfrac{\text{frequency}}{\text{class width}}$

Box and whisker diagrams

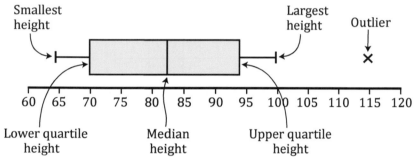

Outliers – values which greatly differ from the norm are shown as crosses on the diagram.

An outlier is any value that is smaller than $Q_1 - 1.5 \times$ IQR or larger than $Q_3 + 1.5 \times$ IQR

Q_1 is the lower quartile
Q_3 is the upper quartile
IQR is the inter-quartile range

Cumulative frequency diagrams

Average spent on a meal out per person (£p)	Frequency	Cumulative frequency
$0 \leq p < 5$	8	8
$5 \leq p < 10$	12	20
$10 \leq p < 15$	22	42
$15 \leq p < 20$	15	57
$20 \leq p < 25$	7	64
$25 \leq p < 30$	5	69
$30 \leq p < 35$	3	72

A cumulative frequency diagram is drawn by plotting the upper class boundary with the cumulative frequency.

The following cumulative frequency diagram has been drawn using the above data – notice the way the median and quartiles are found.

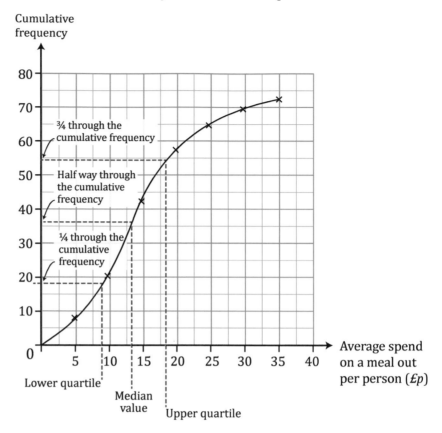

Calculation of the mean

For a **set of values**, mean, $\mu = \dfrac{\sum x_i}{n}$ where $\sum x_i$ is the sum of all the individual values and n is the number of values.

For a **frequency distribution**, mean, $\mu = \dfrac{\sum f x_i}{\sum f}$, where $\sum f x_i$ is the sum of the product of all the values of f multiplied by x and $\sum f$ is the sum of all the frequencies.

Working out the mode – the mode is the value(s) or class that occurs most often. Unlike the mean and median, it is possible to have more than one mode.

Working out the median – the median is the middle value when the data values are put in order of size. The median is at the $\dfrac{n+1}{2}$ value.

Negative, symmetric/no skew and positive skew

The graphs below show the shapes of the distributions along with their corresponding box and whisker diagrams.

Notice the direction of the tails. The direction of the tail for positive skew is in the direction of larger positive values.

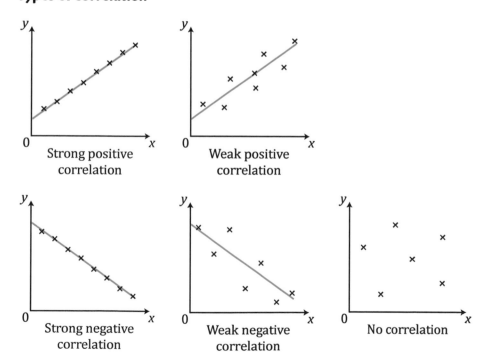

Negatively skew — Q_1, Q_3, Median

Normal (no skew) — Q_1, Q_3, Median

Positively skew — Q_1, Q_3, Median

Scatter diagrams and regression lines

Correlation – a measure of how well two variables are related to each other.

Types of correlation

Strong positive correlation

Weak positive correlation

Strong negative correlation

Weak negative correlation

No correlation

Regression lines

Regression lines are the equations of the lines of best fit and have an equation of the form for that of a straight line $y = mx + c$, where m is the gradient of the line and c is its intercept on the y-axis.

Using regression lines

Interpolation – using the line of best fit within the range of the first and last points.

Extrapolation – using the line of best fit outside the range of the first and last points.

Extrapolation should be used with care. The graph may behave differently outside the range of points plotted.

Measures of central variation (variance, standard deviation, range and interquartile range)

The following measures of the spread of data can be used:

Range – the difference between the largest value and the smallest value in a set of data.

Interquartile range (IQR) – the difference between the upper quartile (Q_3) and the lower quartile (Q_1) so IQR = $Q_3 - Q_1$

$$\textbf{Variance} = \frac{\sum(x_i - \mu)^2}{n}$$

where $\sum(x_i - \mu)^2$ is the sum of the squares of the differences between each value in the set and the mean μ and n is the total number of values.

The **simplified formula for variance** is,

$$\text{variance} = \frac{\sum x_i^2}{n} - \left(\frac{\sum x_i}{n}\right)^2$$

where $\frac{\sum x_i^2}{n}$ is the mean of the squares of the values and $\left(\frac{\sum x_i}{n}\right)^2$ is the square of the mean of the values.

Standard deviation (σ)

The standard deviation (σ) is the square root of the variance

so $$\sigma = \sqrt{\frac{\sum(x_i - \mu)^2}{n}}$$

or $$\sigma = \sqrt{\frac{\sum x_i^2}{n} - \left(\frac{\sum x_i}{n}\right)^2}$$

Looking at exam questions

1   A baker is aware that the pH of his sourdough, y, and the hydration, x, affect the taste and texture of the final product. The hydration is measured in ml of water per 100 g of flour (ml/100 g). The baker researches how the pH of his sourdough changes as the hydration changes. The results of his research are shown in the diagram below:

How pH changes with hydration

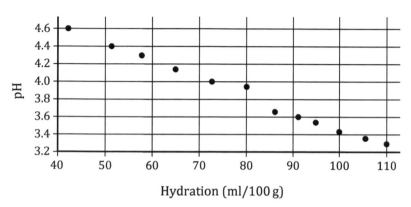

Hydration (ml/100 g)

(a)   Describe the relationship between pH and hydration.   [2]

(b)   The equation of the regression line for y on x is $y = 5.4 - 0.02x$.

   (i)   Interpret the gradient and intercept of the regression line in this context.

   (ii)   Estimate the pH of the sourdough when the hydration is 20 ml/100 g. Comment on the reliability of this estimate.   [4]

Thinking about the question

This question concerns the type of correlation for the line of best fit for a scatter diagram.

It also concerns knowledge of the equation of a straight line in the form $y = mx + c$.

Starting the solution

For part (a) we need to say the type of correlation and whether it is strong or weak and say what it means in the context of the question.

For part (b)(i) we need to look at the units on the y-axis and divide them by the units on the x-axis for the gradient and then work out their significance in the context of the gradient from the equation of the line. The intercept will be the pH at zero hydration.

For part (b)(ii) we need to substitute $x = 20$ into the equation of the line.

11 Data presentation and interpretation

It is best to say how the quantity plotted on the *y*-axis varies with increasing values of the quantity plotted on the *x*-axis.

Think about what the quantity plotted on the *y*-axis divided the quantity plotted on the *x*-axis would represent.

The solution

(a) There is a strong linear relationship with negative correlation showing that the higher the hydration the lower the pH.

(b) (i) The gradient shows that on average, each additional ml of water per 100g of flour decreases the pH by 0.02.

The intercept implies that at zero hydration, the pH would be 5.4.

(ii) $y = 5.4 - 0.02x$

When $x = 20$, $y = 5.4 - 0.02 \times 20$

$y = 5$

The value for x is outside the values of x for the dataset. We are therefore extrapolating to find the value of y and assuming that the line continues with the same equation as before, which it might not do.

2   Basel is a keen learner of languages. He finds a website on which a large number of language tutors offer their services. Basel records the cost, in dollars, of a one hour lesson from a random sample of tutors. He puts the data into a computer program which gives the following summary statistics.

```
Cost per 1 hour lesson
Min       :   10.0
1st Qu. :   16.0
Median  :   17.2
Mean    :   19.8
3rd Qu. :   21.0
Max       :   40.0
```

(a) Showing all calculations, comment on any outliers for the cost of a one-hour lesson with a language tutor.   [4]

(b) Describe the skewness of the data and explain what it means in this context.   [2]

Dafydd is also a keen learner of languages. He takes his own random sample of the cost, in dollars, for a one hour lesson. He produces the following box plot.

Cost in dollars for a one hour lesson

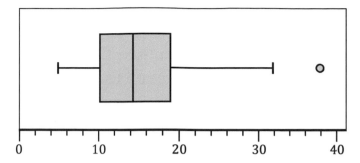

(c) (i) What will happen to the mean if the outlier is removed?

(ii) What will happen to the median if the outlier is removed? [2]

(d) Compare and contrast the distributions of the cost of one-hour language lessons for Dafydd's sample and Basel's sample. [2]

Thinking about the question

Part (a) of the question is about identifying outliers. We need to use the two formulae using the quartiles and the inter-quartile range to indentify the values of any outliers at the lower and higher ends of the distribution. These two formulae will need to be recalled from memory as they are not included in the formula booklet.

For part (b) we have to consider the spread of the data. You can see that the quartiles look as though they are evenly spread from the median. However, more of the data is situated at higher values compared to the median.

For part (c) we need to consider the removal of a higher data value.

For part (d) we need to record some summary statistics using the box and whisker diagram for Dafydd and then make the comparison.

Starting the solution

For part (a) we need to recall the following formulae: Outlier is smaller than $Q_1 - 1.5 \times$ IQR or larger than $Q_3 + 1.5 \times$ IQR.

For part (b) we go by the direction of the tail. As the tail is towards the higher values it will be positive skew. We notice that there is more variation in the costs for the more expensive tutors.

For part (c) if a high outlier is removed it will lower the total of the amounts, which will lower the mean. We need to consider removing the highest value on the median. The median only uses the position of values.

For part (d) we need to compare differences in central tendency (i.e. median) and measures of spread (i.e. range and inter-quartile range) for each of the distributions.

The solution

(a) An outlier is any value that is smaller than $Q_1 - 1.5 \times$ IQR or larger than $Q_3 + 1.5 \times$ IQR

Outlier would be smaller than $16.0 - 1.5 \times (21 - 16) = 8.5$

Outlier would be larger than $21.0 + 1.5 \times (21 - 16) = 28.5$

The max. value of 40 is greater than 28.5 and is therefore an outlier.

As only summary statistics are shown we do not know all the individual data values, so there could be other outliers.

(b) Positive skew.

A few of the tutors are very expensive.

(c) (i) The mean will decrease.

> You would be removing the largest value in the distribution thus lowering the total and hence the mean.

(ii) The median may stay the same or it may decrease.

> The largest value is removed so the median of the remaining numbers will be in the middle of the remaining numbers. This could mean it could move towards lower numbers or the same number.

(d) Any two of the following:

Dafydd's lessons are cheaper on average than Basel's.

Dafydd's lessons are more variable in cost than Basel's.

Both are positively skewed.

Exam practice

1 Data was collected on how the resting heartbeat of a number of people of the same age is affected by the hours of exercise per week and the results of the research are shown in the diagram below:

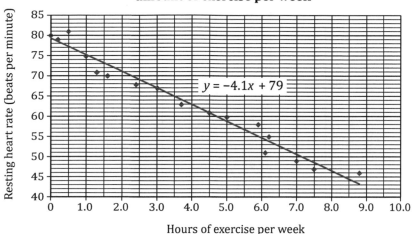

How resting heart rate varies with the amount of exercise per week

$y = -4.1x + 79$

(a) Describe the type of relationship between resting heart rate and the hours of exercise per week. [2]

(b) Using spreadsheet software, the equation of the regression line for y on x was found to be:
$$y = -4.1x + 79$$

   (i) Interpret the gradient of the regression equation. [1]

   (ii) Interpret the intercept of the line with the y-axis. [1]

   (iii) Explain why the spreadsheet software has not drawn the line past the last reading for hours of exercise = 8.8. [1]

   (iv) State whether the relationship between resting heart rate and the hours of exercise per week is likely to be causal. Explain your answer. [1]

   (v) State with a reason, whether the regression line could be used to predict the resting heart rate for an athlete who did on average 5 hours of exercise per day. [1]

2 Megan is a science teacher and she has collected marks from her classes in physics and chemistry and used them to produce the following scatter diagram:

Scatter diagram showing marks in physics and chemistry

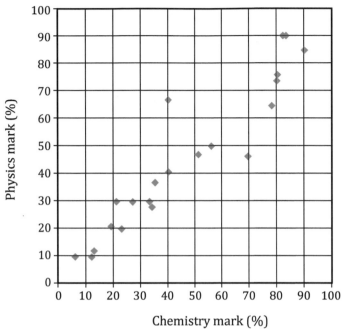

(a) (i) Comment on the correlation between 'Physics marks' and 'Chemistry marks'. [1]

(ii) Interpret the correlation between 'Physics marks' and 'Chemistry marks'. [1]

(b) Megan decides to remove the outlier and then uses software to produce the following table of summary statistics.

	Chemistry mark (%)	Physics mark (%)
Minimum	6.0	10.0
Lower quartile	22.5	26.3
Median	37.5	39.0
Upper quartile	78.5	68.8
Maximum	90.0	90.0

(i) Use the appropriate statistics from this table to prove that the maximum mark for physics is not an outlier. [2]

(ii) Compare and contrast the distribution of physics and chemistry marks. [3]

③ The ages of 30 cars, to the nearest year, traded in at a large new car dealership for the last month are shown here:

1, 1, 1, 1, 2, 2, 2, 2, 2, 2, 2, 3, 3, 3, 3,
3, 4, 4, 4, 4, 5, 5, 6, 6, 7, 8, 8, 9, 10, 20

> Note that the list of ages provided, is in numerical order.

(a) Find the median age. [2]

(b) Find the lower and upper quartiles. [2]

(c) Find any outliers. [1]

④ A local council monitors carbon monoxide levels in the air through a small village. They have collected data to investigate how carbon monoxide levels in parts per million (y) in the atmosphere vary with traffic volume in vehicles per day (x). The results of readings over a series of random days have been shown on the following scatter diagram:

Carbon monoxide levels vs traffic volume

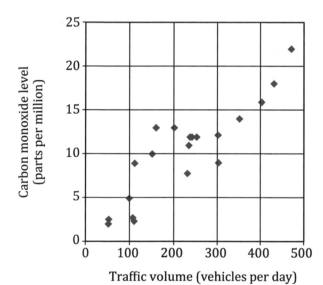

The points shown were used to obtain the linear regression equation of pollution level on traffic volume $y = 0.8x + 1.9$

(a) Describe the correlation between pollution level and traffic volume. [1]

(b) Give an interpretation of the gradient of this regression line. [1]

(c) Give an interpretation of the value 1.9 in the regression equation. [1]

(d) Explain why this model might not be appropriate for predicting the carbon monoxide levels for the air near a very busy motorway. [2]

5 The following box and whisker diagrams show information about students' marks in a chemistry and physics examination. Both exams were marked as a percentage.

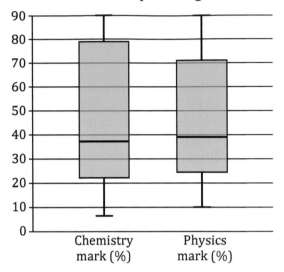

Chemistry mark (%)   Physics mark (%)

(a) Find the interquartile range of marks for each exam. [1]

(b) Using measures of central tendency and measures of spread, compare and contrast the distributions of marks in chemistry and physics. [3]

6 Jacob collects data to see how engine size affects fuel consumption for different cars. He collects pairs of values and uses them to produce the following scatter diagram. The software he uses also adds a regression line and its equation.

Graph to show the variation of fuel consumption with engine size

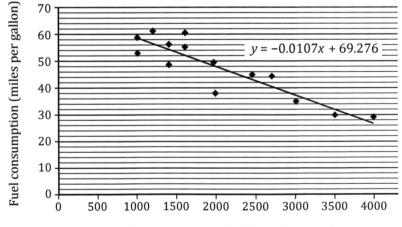

$y = -0.0107x + 69.276$

Fuel consumption (miles per gallon)

Engine capacity (cubic centimetres)

(a) Use the equation to predict the likely fuel consumption for a car with an engine size of 2700 cubic centimetres. [1]
(b) State the type of correlation shown in the above scatter diagram. [1]
(c) State whether the relationship shown by the scatter diagram is causal or non-causal. Explain how you have decided. [1]
(d) Explain the difference between the terms 'extrapolation' and 'interpolation'. [2]
(e) Jacob says he can use the regression line to find the fuel consumption for a car with an engine capacity of 5000 cubic centimetres.
Explain the problem in using this model in this way. [1]

7 A random sample of the time in minutes 30 cars spend in a pay and display car park over a 24-hour period is shown below:

25, 31, 34, 40, 42, 55, 56, 63, 70, 81, 93, 100, 123, 124, 130, 170, 208, 209, 210, 235, 236, 240, 257, 260, 301, 340, 400, 401, 450, 600

The above data is in numerical order.
(a) (i) Determine Q_1 and Q_2 for the above data. [2]
    (ii) Determine if there are any outliers in this set of data. [1]
(b) After a decrease in the cost of parking, the manager of the car park wants to see what effect it has on the distribution of times spent in the car park. The data was collected in the same way as before and the results are as follows:

5, 10, 29, 29, 30, 31, 43, 44, 44, 57, 58, 58, 59, 59, 60, 61, 110, 111, 111, 120, 170, 178, 179, 183, 205, 240, 300, 410, 411, 720

The following statistics were determined using a computer program for all the data:

Minimum	5
Lower quartile	44
Median	60.5
Upper quartile	178.75
Maximum	720
Inter-quartile range	134.75
Mean	137.5

The values 410, 411 and 720 were all found to be outliers. Describe the how the following statistics would change when the three outliers are removed from the set of data.
(i) the inter-quartile range [2]
(ii) the median [2]
(iii) the mean [2]

8 A survey of 50 boys and 50 girls was carried out to find the time spent on social media yesterday. The time spent, t (hours) by each of the boys has been displayed as a cumulative frequency graph and the time spent, t (hours) by each of the girls has been displayed as a box and whisker diagram.
The range for the boys was found to be 8.5 hours.
Compare and contrast the times spent by the boys and the girls.

Cumulative frequency graph for boys showing how much time in hours was spent yeaterday on social media

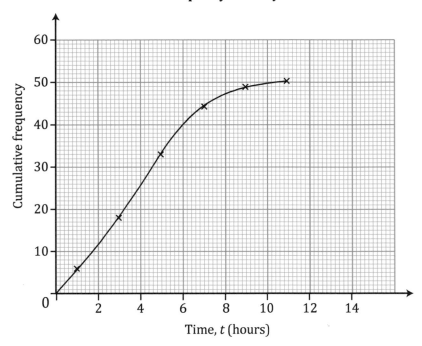

Box and whisker diagram for girls showing how much time in hours was spent yesterday on social media

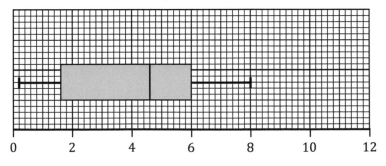

[5]

Cumulative frequency graph for boys showing how much time in hours were spent yesterday on social media.

Box and whisker diagram for girls showing how much time in hours were spent yesterday on social media.

9 An airline wants to monitor the weight of hand luggage of its passengers. The weights (x) of hand luggage are recorded for 20 passengers and the following set of summary values is obtained:

$$\sum x_i = 92$$

$$\sum x_i^2 = 476$$

Calculate the variance and standard deviation for this set of data and give both answers correct to two significant figures. [3]

10 Data was collected concerning the amount of carbohydrate and calories in breakfast cereal. The following scatter diagram was produced:

How calories vary with carbohydrate in cereal

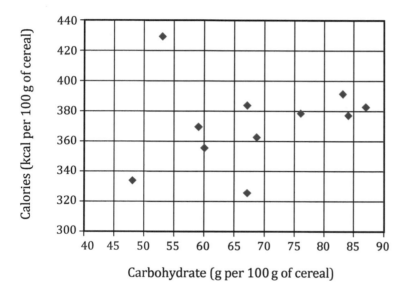

Carbohydrate (g per 100 g of cereal)

(a) Describe the correlation shown by the above diagram in context. [2]

(b) The investigator has said that this diagram overemphasises the increase in calories for an increase in carbohydrate. Explain why they have said this. [1]

(c) Do you think the relationship between calories and carbohydrate is causal? Explain your answer. [2]

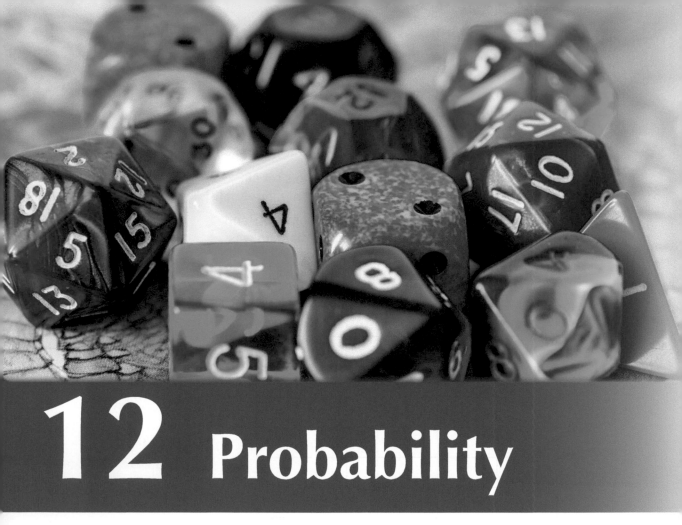

12 Probability

Prior knowledge

You will need to make sure you fully understand the following from your GCSE studies:

- The probability scale from 0 to 1
- The probability of an event not occurring is one minus the probability of it occurring
- Identifying all outcomes from two events using tree or Venn diagrams

- Recognise the difference between mutually exclusive and independent events
- Calculations of probabilities for mutually exclusive and independent events
- Conditional probabilities and the multiplication law for dependent events

Quick revision

Venn diagrams

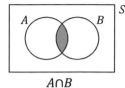

$A \cap B$

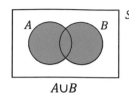

$A \cup B$

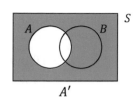

A'

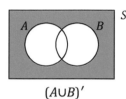

$(A \cup B)'$

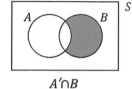

$A' \cap B$

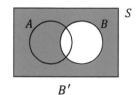

B'

Complementary events

The probability that A does not occur (i.e. A') is one minus the probability that event A does occur.

$$P(A') = 1 - P(A)$$

The meaning of independent events and mutually exclusive events

When an event has no effect on another event, they are said to be independent events. For example, if event A occurs then it will have no effect on event B happening and vice versa.

If events A and B are mutually exclusive then A or B can occur but not both.

The addition law for mutually exclusive events

$$P(A \cup B) = P(A) + P(B)$$

The generalised addition law

$$P(A \cup B) = P(A) + P(B) - P(A \cap B)$$

Multiplication law for independent events

$$P(A \cap B) = P(A) \times P(B)$$

Note that this formula applies only to independent events.

107

Looking at exam questions

1 Branwen performs a survey about ownership of dogs, cats and birds as pets.

She finds the following after surveying a group of 60 people she knows:

> 10 own cats only.
>
> 17 own dogs only.
>
> 5 own birds only.
>
> 7 own cats and dogs.
>
> 7 own cats and birds.
>
> 2 own birds and dogs only.
>
> 15 do not own a pet.

(a) Find the number of people who have all three types of pet. [5]

(b) Find the probability that a person chosen at random owns all 3 pets. [1]

(c) Find the probability that a person chosen at random from the group owns a cat and a bird only. [1]

(d) Find the probability that if a person owns a cat, they also own a dog. [2]

Thinking about the question

This is a question on probability, and it is best solved by first drawing a Venn diagram. A quick look through the question indicates there are 3 areas with each area representing the ownership of the different pets. It should be noted that we are interested in the ownership of the pets, so the numbers refer to the numbers of people owning the pets and not the number of pets themselves.

Starting the solution

For part (a) we need to let the number of people who own all three types of pet equal x. We can then work from the centre outwards marking in the various areas either the numbers (if they are known) or the algebraic expression. We need to remember to mark the people who do not own one or more of the pets outside the circles.

We can then add up the numbers or expressions for all the areas and then equate them to the total number of people which is 60.

For parts (b) and (c) the probabilities will be out of the whole group (i.e. out of 60). We need to remember to cancel any fractions if possible.

For part (d) we know the person owns a cat so the denominator for the fraction representing the probability will be the number who own a cat.

Solution

(a) Let the number of people who have all three pets = x.

Number of people who have cats and dogs only = $7 - x$.

Number of people who have cats and birds only = $7 - x$.

We can now fill in the rest of the numbers in the relevant areas.

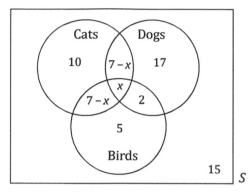

Total number of people = 60

Hence,

$$10 + 7 - x + 17 + 7 - x + x + 2 + 5 + 15 = 60$$

$$63 - x = 60$$

$$x = 3$$

Number of people with all three types of pet = 3.

> The numbers in all the areas in the Venn diagram are added together and then equated to 60.

(b) Probability of owning all three types of pet = $\dfrac{3}{60} = \dfrac{1}{20}$

(c) Probability of owning a cat and bird only = $\dfrac{4}{60} = \dfrac{1}{15}$

> The number owning a cat and a bird only = $7 - x = 7 - 3 = 4$.

(d) Number of people owning cats = $4 + 3 + 4 + 10 = 21$

Number owning cats and dogs = 7

Probability that if a person owns a cat,

they also own a dog = $\dfrac{7}{21} = \dfrac{1}{3}$.

2   The Venn diagram shows the subjects studied by 40 sixth form students. F represents the set of students who study French, M represents the set of students who study mathematics and D represents the set of students who study drama. The diagram shows the number of students in each set.

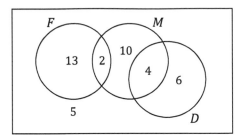

(a)   Explain what $M \cap D'$ means in this context.   [1]

(b)   One of these students is chosen at random. Find the probability that this student studies:

(i)   Exactly two of these subjects,

(ii)   Mathematics or French or both.   [3]

(c)   Determine whether studying mathematics and studying Drama are statistically independent for these students.   [3]

Thinking about the question

This is a Venn diagram question and you can see that all the various parts of the diagram are known. It is important to note that the numbers refer to the number of students in the sets and the sets themselves represent the actual students.

Starting the solution

For part (a) we are looking for the region outside of set D but is overlapped with set M. We have to explain this region in words.

For part (b)(i) we notice there are two such regions, so we need to find $P(F \cap M)$ and add it to $P(M \cap D)$.

For part (b)(ii) we need to first find the total of the number taking maths only, French only, maths and French, and maths and drama, and express it as a fraction of the total number of students.

For part (c) we need to see if $P(M) \times P(D) = P(M \cap D)$. If this is true we have proved independence.

Solution

(a) $M \cap D'$ represents the set of students who study mathematics but not drama.

> Do not write that is the number of students (or the probability) who study mathematics but not drama.

(b) (i) P(exactly two subjects) = $P(F \cap M) + P(M \cap D)$

$$= \frac{2}{40} + \frac{4}{40} = \frac{6}{40} = \frac{3}{20}$$

(ii) P(Mathematics or French or both) $= \dfrac{4 + 10 + 2 + 13}{40} = \dfrac{29}{40}$

(c) For statistical independence, we need to show that:

$$P(\text{Maths}) \times P(\text{Drama}) = P(\text{Maths and drama})$$

$$P(M) \times P(D) = P(M \cap D)$$

Now $\qquad P(M) = \dfrac{16}{40} \qquad$ and $\qquad P(D) = \dfrac{10}{40}$

$$P(M) \times P(D) = \frac{16}{40} \times \frac{10}{40} = \frac{1}{10} \text{ and } P(M \cap D) = \frac{4}{40} = \frac{1}{10}$$

Hence $P(M) \times P(D) = P(M \cap D)$ so they are statistically independent events.

Exam practice ◀◀◀◀ ◀◀

① The events A and B are independent such that $P(A) = 0.7$ and $P(B) = 0.4$.
(a) Find $P(A \cup B)$. [3]
(b) Find the probability that:
    (i) Exactly one of A and B will occur,
    (ii) Neither A nor B will occur. [6]

② The events A and B are such that $P(A) = 0.5$ and $P(A \cup B) = 0.7$.
Determine the value of $P(B)$ in each of the cases when:
(a) A and B are mutually exclusive, [2]
(b) A and B are independent, [4]

③ The two independent events A and B are such that $P(A) = 0.2$, $P(A \cup B) = 0.4$.
(a) Evaluate $P(B)$. [4]
(b) Find the probability that exactly one of the two events occurs. [3]
(c) Given that exactly one of the two events occurs, calculate the probability that A occurs. [3]

4 The independent events A and B are such that:
$P(A) = 0.6$, $P(B) = 0.3$.
Find the value of:
(a) $P(A \cup B)$, [3]
(b) $P(A \cup B')$. [3]

4 The events A and B are independent such that $P(A) = 0.7$ and $P(B) = 0.4$.
(a) Find $P(A \cup B)$. [3]
(b) Find the probability that:
    (i) Exactly one of A and B will occur,
    (ii) Neither A nor B will occur. [6]

4 The events A, B are such that $P(A) = 0.4$, $P(B) = 0.2$. Determine the value of $P(A \cup B)$ when:
(a) A, B are mutually exclusive, [2]
(b) A, B are independent, [3]
(c) $B \subset A$. [1]

4 In a class of 30 students, 12 are studying French, 15 are studying Spanish and 8 are studying neither French nor Spanish. A student is selected at random from this class.
(a) Find the probability that the student is studying both French and Spanish. [4]
(b) Find the probability that the student is studying French but not Spanish. [2]

4 Students in a school can take one or two of French (F), German (G) and Spanish (S) at a school. Some students do not take a language. This information is shown by the following Venn diagram.

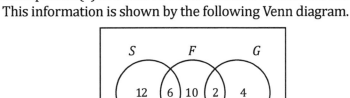

(a) Explain what $(S \cup F \cup G)'$ means in this context. [1]
(b) One of the students is chosen at random. Find the probability that this student studies:
    (i) French only
    (ii) French or German or both. [3]
(c) Determine whether or not taking Spanish or French is statistically independent. [2]

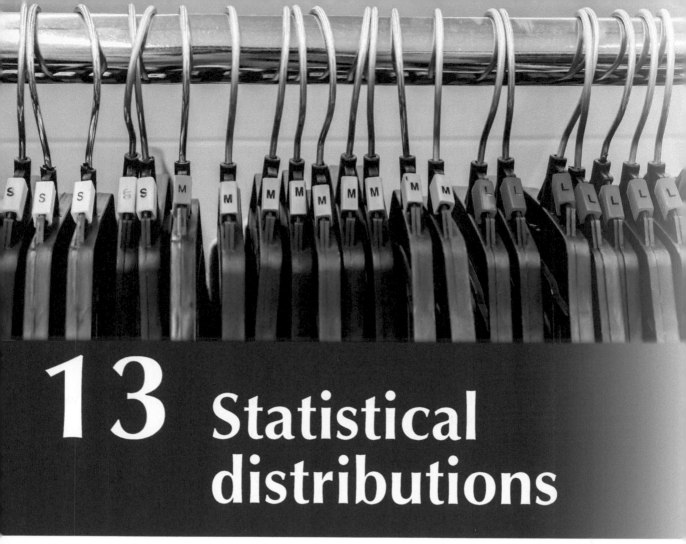

13 Statistical distributions

Prior knowledge

You will need to make sure you fully understand the following from your GCSE studies:

- Substitution of numbers into formulae

- Use of the probability of an event not occurring is one minus the probability that it occurs

Quick revision

Using statistical distributions

Statistical distributions can be used to work out probabilities. The binomial distribution and Poisson distribution can be used to work out the probability of obtaining a random variable called X, less than or equal to a certain value, x (written as $P(X \leq x)$). The random variable X does not have a fixed value but instead can take a range.

The binomial distribution

> This formula is included in the formula booklet.

For a fixed number of trials, n, each with a probability p of occurring, the probability of x successes is given by the formula:

$$P(X = x) = \binom{n}{x} p^x (1 - p)^{n-x}$$

Example

A fair coin is tossed ten times. What is the probability that it lands on heads five times?

> To find $\binom{10}{5}$ on a calculator we use the nCr button. You first enter n, 10 in this case, then press shift and then the nCr button and finally enter the value of r, which is 5 in this case.

Here $n = 10$, $p = 0.5$ (i.e. the probability it landed on heads) and we require $P(X = 5)$.

$$P(X = x) = \binom{n}{x} p^x (1 - p)^{n-x}$$

$$P(X = 5) = \binom{10}{5} 0.5^5 (1 - 0.5)^{10-5}$$

$$= 0.2461$$

Finding $P(X \leq x)$ using the binomial (CD) distribution function on a calculator

1   Use the binomial CD function on your calculator.

2   Enter your values for x, N and p.

3   The result is $P(X \leq x)$.

Assumptions that must be made when using the binomial probability distribution as a model

● Independent trials – the probability of one event happening must not depend on any other event happening. This means the probability of an event occurring stays the same.

● There must be only success and failure – we must be able to classify the event into it happens or it doesn't.

> If you don't have a fixed number of trials, then another probability distribution such as the Poisson distribution will have to be used.

Use the binomial distribution when

● There are a fixed number of trials (i.e. the value of n is known).

● The probability of success or failure, p, is known and remains constant.

The Poisson distribution

In a particular interval (i.e. time or space), the probability of an event X occurring x times is given by the formula:

$$P(X = x) = e^{-\lambda} \frac{\lambda^x}{x!}$$

where $\lambda = \mu = E(X)$ and $x = 0, 1, 2, 3, 4, \ldots$

Note that μ is the mean value and $E(X)$ is the expected value.

This is in the formula booklet, so you don't need to remember it.

Assumptions that must be made when using the Poisson probability distribution as a model

- Events occur independently, so if an event occurs, it does not affect the probability of another event occurring in the same time period.

- The rate of occurrence of an event is constant.

Use the Poisson distribution when

- You want to find the probability of an event happening in a given time slot or space slot.

- You know the rate at which the event occurs.

The number of patients visiting a GP in a day or the number of defects per metre in a sheet of material, are discrete variables that follow a Poisson distribution.

Finding $P(X \leq x)$ using the Poisson CD function on a calculator

1   Use the Poisson CD function on your calculator.

2   Enter your values for x and λ.

3   The result is $P(X \leq x)$.

The CD means 'cumulative distribution' as all the probabilities for values up to and including the value of x you have entered.

The discrete uniform distribution

This is a distribution where all the outcomes are equally likely. If there are N possible outcomes, the probability of a particular outcome $= \dfrac{1}{N}$.

Understanding the language used in the questions

Certain phrases appear in questions and you must be clear about what they mean. Here are a few example phrases and what they mean:

Suppose X is a random variable and each of the following statements is used to describe the range of values X can take.

Phrase	Meaning
At least 5	$X \geq 5$
No more than 8	$X \leq 8$
More than 10	$X > 10$
Less than 11	$X < 11$
At most 12	$X \leq 12$

Looking at exam questions

1   The random variable X has the binomial distribution B(16, 0.3).
    Showing your calculation, find $P(X = 7)$.   [2]

Thinking about the question

This is a straightforward question using the binomial distribution.
B(16, 0.3) means a binomial distribution with a total number of trials,
$n = 16$ and a probability of occurring, $p = 0.3$. Notice the 'Showing your
calculation' wording in the question – you need to make it clear what
values you are using in your calculation so just the numerical answer
would not be sufficient.

Starting the solution

We need to obtain the formula for $P(X = x)$ for a binomial distribution
with $p = 0.3$, $x = 7$ and $n = 16$. The formula is looked up in the formula
booklet.

> Do not put the values
> straight into the
> calculator without
> showing any working.
> You must show your
> calculation as the
> question asks you to
> do this.

The solution

$$P(X = x) = \binom{n}{x}p^x (1 - p)^{n-x}$$

$$P(X = 7) = \binom{16}{7}(0.3)^7 (1 - 0.3)^{16-7}$$

$$= \binom{16}{7}(0.3)^7 (0.7)^9$$

$$= 0.10096...$$

Alternative solution

> You can use the
> binomial CD function
> on the calculator with
> $N = 16$, $p = 0.3$ and
> $x = 7$ and 6 respectively
> to find $P(X \le 7)$ and
> $P(X \le 6)$. These are
> then subtracted to
> give the required
> probability.

You could have found $P(X = 7)$ by using:

$$P(X = 7) = P(X \le 7) - P(X \le 6)$$

$$= 0.9256 ... - 0.8246 ...$$

$$= 0.10096 ...$$

2   The probability of a bird's egg hatching is 0.5 and the probability
    is independent of all other eggs hatching. A bird lays n eggs
    and a statistician calculates that the probability of all the eggs
    hatching is 0.0078. Find the value of n.   [4]

Thinking about the question

As the exact probability is known, we use the binomial distribution
to answer this question. The binomial formula is obtained from the
formula booklet.

Thinking about the solution

Since we want all the eggs to hatch, it means the value of x is n as well as the total number of eggs being n. The probability of each egg hatching is 0.5.

All these values are entered into the formula and equated to the probability of 0.0078 and we can then look at the resulting equation to see how the value of n can be found.

The solution

Let the random variable X be the number of eggs hatched.

$$X \sim B(n, p)$$

$$X \sim B(n, 0.5)$$

$$P(X = x) = \binom{n}{x} p^x (1 - p)^{n - x}$$

In this case $x = n$ so we have

$$P(X = n) = \binom{n}{n} p^n (1 - p)^{n - n}$$

$$= \binom{n}{n} 0.5^n (1 - 0.5)^0$$

$$= 0.5^n$$

Now $\qquad\qquad 0.5^n = 0.0078$

Taking logs of both sides

$$\log 0.5^n = \log 0.0078$$

$$n \log 0.5 = \log 0.0078$$

$$n = \frac{\log 0.0078}{\log 0.5} = 7$$

> This is binomial as the exact probability of an egg hatching is known.

> Note that $\binom{n}{n} = 1$ and $0.5^0 = 1$.

> Equations involving powers like this can be solved by taking logs of both sides. Look back at your Pure Maths notes if you are unsure about this.

> **Common mistake**
> This is not the same as $\log \left(\frac{0.0078}{0.5} \right)$.

---

3 Naomi produces oak tabletops, each of area 4.8 m². Defects in the oak tabletops occur randomly at a rate of 0.25 per m².

(a) Find the probability that a randomly chosen tabletop will contain at most 2 defects. [3]

(b) Find the probability that, in a random sample of 7 tabletops, exactly 4 will contain at most 2 defects each. [3]

Thinking about the question

Part (a) is a question about the Poisson distribution as Poisson distributions are used to model the number of events (i.e. defects in this case) occurring within a given interval of time or space (i.e. the space in this case is the area).

From part (a) you will know the probability p so part (b) can be worked out using a binomial distribution.

Thinking about the solution

For part (a) we need to work out the mean number of defects in an area of 4.8 m^2 and then use this with the Poisson formula obtained from the formula booklet, to work out the probability.

For part (b) we need to obtain the formula for the binomial distribution and use the probability p obtained from part (a) and tables to find the required probability.

The solution

(a)   There are 0.25 defects in an area of 1 m^2, in an area of 4.8 m^2 there would be a mean number of defects = 0.25 × 4.8 = 1.2.

Mean defects, $\lambda = 1.2$

Let the random variable X be the number of defects per tabletop.

$$X \sim \text{Po}(\lambda)$$

$$X \sim \text{Po}(1.2)$$

> This is in the formula booklet so you don't need to remember it.

The Poisson formula is looked up: $P(X = x) = e^{-\lambda}\dfrac{\lambda^x}{x!}$

We want to find $P(X \le 2)$

Now $P(X \le 2) = P(X = 0) + P(X = 1) + P(X = 2)$

$$P(X = 0) = e^{-1.2}\frac{1.2^0}{0!} = 0.3012$$

> Note you could have used the Poisson cumulative distribution function on your calculator to solve this problem. All you need to do is enter $x = 2$ and $\lambda = 1.2$ and then = and the calculator works out the answer.

$$P(X = 1) = e^{-1.2}\frac{1.2^1}{1!} = 0.3614$$

$$P(X = 2) = e^{-1.2}\frac{1.2^2}{2!} = 0.2169$$

$$P(X \le 2) = 0.3012 + 0.3614 + 0.2169 = 0.8795$$

(b)   Let the random variable Y be the number of 4.8 m^2 tables containing at most 2 defects.

$$Y \sim \text{B}(n, p)$$

$$Y \sim \text{B}(7, 0.8795)$$

> Notice we have changed the random variable to Y so that we don't get it mixed up with part (a).

$$P(X = x) = \binom{n}{x}p^x (1 - p)^{n-x}$$

> Notice that the probability we use here is the value obtained from the answer to part (a).

$$P(Y = 4) = \binom{7}{4}0.8795^4 (1 - 0.8795)^{7-4}$$

$$= 0.0366$$

Exam practice

1 The probability that a machine part fails in its first year is 0.05 independently of all other parts. In a batch of 20 randomly selected parts, find the probability that in the first year that:
  (a) exactly one part fails,   [4]
  (b) more than 4 parts fail.   [2]

2 (a) The number of cars sold per week by a car dealer can be modelled as a random variable with mean 6. Find the probability that, during a randomly chosen week, the dealer sells:
      (i)   at least 4 cars,
      (ii) exactly 6 cars.   [5]

  (b) The dealer also sells motorbikes. Assuming that the number of motorbikes sold per week can also be modelled as a random variable but with mean 1.12, find the probability that, in a randomly chosen week, the dealer sells:
      (i)   exactly 2 motorbikes,
      (ii) at least 2 motorbikes.   [5]

3 A survey is being conducted on the number of newts living in a pond. The pond is a rectangle of length 6 m and width 4 m. It is assumed that newts can be found in the pond at a rate of 0.5 per m^2 of surface area.
  (a) Give one reason why the number of newts can be modelled using a Poisson distribution.   [1]
  (b) Find the probability, that in a randomly chosen surface area of 2 m^2 of this pond, you would find at most 3 newts.   [3]
  (c) Find the probability, that in a random sample of four 2 m^2 randomly chosen areas, two of the areas would contain 2 newts each.   [3]

4 An online booking form for a holiday consists of 4 pages. When users enter their details they make on average 2 mistakes per page.
  (a) Explain why a binomial distribution would be an inappropriate model to use for this situation.   [1]
  (b) Using a Poisson distribution:
      (i)   Find the probability that a page of the form chosen at random has no mistakes on it.   [2]
      (ii) The probability that a page of the form chosen at random has four mistakes on it.   [2]

5 When Alan types a report, the number of errors on each page has a Poisson distribution with mean 0.95, independently of all other pages.
  (a) Find the probability that a randomly selected page contains:
     (i) no errors,
     (ii) either 3 or 4 errors.   [3]
  (b) Alan types a 4-page report. Calculate the probability that:
     (i) There are no errors anywhere in the report.
     (ii) The first error occurs on the third page.   [5]

6 The number of accidents per week at a certain roundabout has a distribution with mean 2.75.
  (a) Find the probability that, during a randomly chosen week, the number of accidents at this roundabout is:
     (i) exactly 4,
     (ii) more than 2.   [5]
  (b) Find the probability that in three randomly picked weeks there are no accidents, giving your answer to one significant figure.   [3]

7 A golfer is practising short putts. Jessie knows from past experience that the probability of getting the ball in the hole with one putt is 0.7.
  Jessie tries 10 practice putts.
  (a) Find the probability that Jessie holes the ball with one putt five or more times.   [2]
  (b) Jessie tries 10 practice putts on 4 different occasions. What is the probability that she will hole the ball in one shot five or more times on each of the four occasions?   [2]
  (c) State one assumption you have made when modelling the situation using your chosen distribution.   [1]

8 The number of emails, X, arriving per day can be modelled by a Poisson distribution with mean 24.
  (a) Explain why a Poisson distribution can be used to produce a model of this situation in order to work out the probabilities of numbers of emails.   [1]
  (b) Calculate:
     (i) $P(X = 20)$.
     (ii) $P(X \leq 24)$.   [5]
  (c) Determine $P(30 \leq X \leq 40)$.   [3]

9 Customers randomly arrive at the returns desk of a large store at a mean rate of 20 per hour.
   (a) Explain why it is appropriate to model the number of customers arriving at the desk in a certain time as a Poisson distribution. [1]
   (b) Jane is one of four staff who works the returns desk. Assuming that customers spread themselves equally between the desks, find the probability that in a 30-minute period, Jane will deal with 4 customer returns. [3]
   (c) Find the time interval in minutes for which the probability of Jane dealing with more than 5 customer returns is approximately 0.1. [3]

10 James, a professional darts player, can hit the bullseye with a probability of 0.2 for each dart he throws.
   (a) Name the probability distribution that can be used to model the number of bullseyes obtained when a certain number of darts are thrown. [1]
   (b) State two assumptions that must be made in order to use the probability distribution you have named in part (a). [2]
   (c) Find the probability that James hits the bullseye at least three times with 10 throws of a dart. [3]
   (d) James throws ten darts on three different occasions. Find the probability that he hits the bullseye on each of these separate occasions at least three times each. [2]

14 Statistical hypothesis testing

Prior knowledge

You will need to make sure you fully understand the following from your GCSE studies:

- Specifying hypotheses

Quick revision

Hypothesis testing

Hypothesis testing is used to test a hypothesis about the probability of the number of times (X) a certain property, called the test statistic, occurs.

The test statistic X, is modelled by a binomial distribution $B(n, p)$ where p is the probability of the event occurring in one trial and n is the total number of trials.

Null and alternative hypotheses

Null hypothesis $H_0 : p$ = a value.

Alternative hypothesis for a one-tailed test is $\quad H_1 : p <$ a value

$\qquad\qquad\qquad\qquad\qquad$ or $\quad H_1 : p >$ a value.

Alternative hypothesis for a two-tailed test is $\quad H_1 : p \neq$ a value.

The null hypothesis is always the status quo (i.e. nothing has changed).

Significance level

Assuming the null hypothesis is true, the significance level (α) indicates how unlikely a test value x, needs to be before the null hypothesis H_0 is rejected.

The significance level for this topic can be 1% ($\alpha = 0.01$), 5% ($\alpha = 0.05$) or 10% ($\alpha = 0.1$).

The critical value and critical region

Here you use the significance level along with the values of n and p to find the set of values of X that would cause the null hypothesis to be rejected.

In all of the following cases the significance level is taken as 5% but this can be changed to 1% or 10% or any other significance level.

If H_0 contains a < sign, the critical region will be the lower 5% of values of X (i.e. in the lower tail). This critical value, a, is the first value of the test statistic having a probability of less than or equal to 0.05 of occurring and the critical region will be $X \leq a$.

If H_0 contains a > sign the critical region will be the upper 5% of values of X (i.e. in the upper tail) so we find the first value of X which has a probability equal to or greater than 0.95. This value of X is the critical value, a, and the critical region will be $X \geq a$.

If H_0 contains a $\neq$ sign the critical region will be the upper 2.5% and lower 2.5% of values of X (i.e. in the upper tail and lower tails). We use the techniques outlined in the previous two paragraphs to find the two critical values and the critical regions in each of the two tails.

Type I and Type II errors

Type I error - A type I error is made when you incorrectly reject a true null hypothesis.

Type II error - A type II error is made when you fail to reject a false null hypothesis.

p-values

The *p*-value is the probability that the observed result or a more extreme one will occur under the null hypothesis.

Assuming that the probability distribution of *X* is $B(n, p)$ the probability of obtaining a value (called the *p*-value) of the test statistic or a more extreme one can be found using tables, or preferably a calculator capable of calculating statistical distributions. If this value is less than or equal to the level of significance, the null hypothesis is rejected.

Example – a one-tailed test at the lower tail of the distribution

The chances of a washing machine having a fault in its first year of use is $\frac{1}{10}$. A new model has been introduced which is supposed to be an improvement. The engineers have conducted tests and these revealed that there would be faults in the first year with 1 out of 35 machines.

The engineers say that this is an improvement.

Conduct a hypothesis test at the 5% level of significance and state your conclusion in context.   [6]

Answer

If *p* is the probability of a machine being faulty in its first year.

Null hypothesis is	$H_0 : p = 0.1$	Make sure you define what both *p* and *X* are.
Alternative hypothesis is	$H_1 : p < 0.1$	

> Before you start, you need to think whether you need a one- or two-tailed test. As you are looking at the just the lower end of the distribution, this is a one-tailed test.

If the machines are more reliable then the probability of a machine being faulty will be less than 0.1. This is a one-tailed test at the lower end of the distribution.

The random variable, *X*, is the number of faulty machines in the first year.

> This says you are modelling the situation using a binomial distribution.

Assuming the null hypothesis is true, $X \sim B(n, p)$.

$$X \sim B(35, 0.1).$$

$$P(X \leq 1) = 0.12$$

> We can use the binomial CD function on the calculator with $x = 1$, $N = 35$ and $p = 0.1$ to work out $P(X \leq 1)$.

This probability is the *p*-value and it is the probability that 1 or 0 (i.e. ≤ 1) machines out of the 35 will develop a fault. This value is then compared with the significance level which is 0.05 in this question. If the *p*-value is less than or equal to 0.05 then the null hypothesis should be rejected otherwise there is not enough evidence to reject it.

0.12 > 0.05, so as the *p*-value is greater than the significance level, we conclude that there is not enough evidence to reject the null hypothesis.

Hence there is evidence that the new model is not more reliable than the older model.

Example – a one-tailed test at the upper tail of the distribution

Joshua takes cuttings of geranium plants. Not all the cuttings successfully grow into plants. His past success rate for these cuttings has been 70%.

There is a new rooting powder on the market and the company says that it encourages more cuttings to root.

Joshua decides to test the claim and he takes 20 cuttings and 18 successfully take root and grow.

(a) Write down suitable null and alternative hypotheses he could use to test the claim made by the company. [2]

(b) At the 5% level of significance, test to see if the claim by the company is correct. [6]

Answer

(a) If *p* is the probability of a cutting successfully growing.

Null hypothesis is $H_0 : p = 0.7$

Alternative hypothesis is $H_1 : p > 0.7$

(b) $P(X \geq 18) = 1 - P(X < 18) = 1 - P(X \leq 17)$

We now use the binomial CD on the calculator (or you could use tables) with $x = 17$, $p = 0.7$ and $N = 20$.

$$P(X \leq 17) = 0.9645$$

$$P(X \geq 18) = 1 - 0.9645$$

$$= 0.0355$$

Now $0.0355 < 0.05$

Hence the null hypothesis is rejected as there is sufficient evidence, at the 5% level of significance, to conclude that the powder has increased the success rate with the cuttings.

You must state the conclusion in context. Don't just say that H_0 is rejected. Also do not say that you accept the alternative hypothesis H_1.

Example – a two-tailed test

A coin is tossed 50 times and it lands on heads 35 times. Using a significance level of 5% and a two-tailed hypothesis test, determine whether there is sufficient evidence to conclude that the coin is biased. [6]

Ensure that your conclusion is made in the context of the question. Don't just say we 'fail to reject the null hypothesis'.

The rooting powder is supposed to increase the probability of rooting.

Watch out

Make sure you use this format for specifying the hypotheses.

This is the *p*-value which is now compared with the level of significance of 0.05.

The *p*-value is less than the significance level so the null hypothesis is rejected.

We need to test for bias – we don't know whether the coin is biased for heads or tails so we need to perform a two-tailed test.

You will always have a ≠ in the null hypothesis for a two-tailed test.

The expected number of heads you would expect in 50 tosses is 25 heads. The number of heads (35) is greater than this so we need to look at the upper tail (i.e. X ≥ 35).

For a two-tailed test we need to halve the significance level (0.05) to give 0.025 before comparing it with the p-value.

Answer

If p is the probability of a head being tossed.

Null hypothesis is $\qquad$ $H_0 : p = 0.5$

Alternative hypothesis is $\qquad$ $H_1 : p \neq 0.5$

The random variable, X, is the number of heads obtained.

Assuming the null hypothesis is true, $\qquad X \sim B(n, p)$

$$X \sim B(50, 0.5).$$

$$P(X \geq 35) = 1 - P(X < 35) = 1 - P(X \leq 34)$$

$$= 1 - 0.9967 = 0.0033 \qquad \text{(this is the p-value)}$$

0.0033 < 0.025, so there is sufficient evidence to reject the null hypothesis and conclude that there is evidence that the coin is biased.

Example – finding critical values

A random variable X has a binomial distribution $B(30, 0.2)$. A hypothesis test is used to test $H_0 : p = 0.2$ against $H_1 : p \neq 0.2$ using a significance level of 10%.

Find the critical values and critical regions for this test. [7]

Answer

As this is a two-tailed test, the probability in each tail will be 5% (i.e. 0.05, half of the significance level).

Considering the lower tail – we need to find the first value of X where the probability is 0.05 or less.

Using the binomial CD function on the calculator with $N = 30$ and $p = 0.2$ we try different values of X in turn like this:

$P(X \leq 3) = 0.1227$ (a starting value for X shows we need to try a lower value for X as the probability is too high).

$P(X \leq 2) = 0.0442$ (this is below 0.05).

Hence the critical value in the lower tail is $X = 2$ and the critical region is $X \leq 2$.

Considering the upper tail we need to find the first probability of $(1 - 0.05) = 0.95$ or higher.

Notice that P(X ≥ 11) is 0.0256 so as this is the first value going from smaller to larger values of X and as 0.0256 < 0.05 this value of X is the critical value.

$P(X \leq 10) = 0.9744$
so $\quad P(X > 10) = P(X \geq 11) = 1 - P(X \leq 10) = 1 - 0.9744 = 0.0256$

$P(X \leq 9) = 0.9389$
so $\quad P(X > 9) = P(X \geq 10) = 1 - P(X \leq 9) = 1 - 0.9389 = 0.0611$

Hence $X = 11$ is the critical value and $X \geq 11$ is the critical region.

Hence the critical values are 2 and 11 and the critical regions are $X \leq 2$ and $X \geq 11$.

Looking at exam questions

1   Edward can correctly identify 20% of types of wild flower. He studies some books to see if he can improve how often he can correctly identify types of wild flower. He collects a random sample of 10 types of wild flower in order to test whether or not he has improved.

(a) (i)   Write suitable hypotheses for this test.

(ii) State a suitable test statistic that he could use.   [2]

(b) Using a 5% level of significance, find the critical region for this test.   [3]

(c) State the probability of a Type I error for this test and explain what it means in this context.   [2]

(d) Edward correctly identifies 4 of the 10 types of wild flower he collected. What conclusion should Edward reach?   [2]

Thinking about the question

This is a standard hypothesis question with plenty of scaffolding to guide you through the steps.

Starting the solution

For part (a) remember that the null hypothesis is the status quo (i.e. the situation before any improvement in identification). The random variable is X so a sentence is needed to explain what X represents.

For part (b) as we are looking to see if there is an improvement in identification, we need to look at the upper tail of the distribution. Using the binomial CD function on the calculator we need to find the smallest value of X for which the probability exceeds 0.95 and then add one to it to find the critical value. The critical region will be all the values of X greater than or equal to the critical value.

For part (c) we need to find the probability of rejecting the null hypothesis.

The solution

(a) (i)   If p is the probability of Edward correctly identifying types of wild flower:

Null hypothesis is $H_0 : p = 0.2$

Alternative hypothesis is $H_1 : p > 0.2$

(ii) The test statistic, X, is the number of times he correctly identifies a type of wild flower from the 10 types of wild flower.

As the significance level is 0.05, 0.0328 is less than this so $X = 5$ is the critical value and the critical region is $X \geq 5$.

(b) Assuming the null hypothesis is true, $X \sim B(10, 0.2)$.

Using the binomial CD function on the calculator with $N = 10$ and $p = 0.2$ we try different values of X in turn like this:

$P(X \leq 3) = 0.8791$ so $P(X \geq 4) = 1 - 0.8791 = 0.1209$

$P(X \leq 4) = 0.9672$ so $P(X \geq 5) = 1 - 0.9672 = 0.0328$

$P(X \leq 5) = 0.9936$ so $P(X \geq 6) = 1 - 0.9936 = 0.0064$

The first probability that exceeds 0.95 is 0.9672 which corresponds to an X value of 4.

The critical value will be one greater (i.e. $X = 5$) and the critical region is $X \geq 5$.

(c) The probability of a type I error $= P(X \geq 5) = 1 - P(X \leq 4)$

$$= 1 - 0.9672$$

$$= 0.0328$$

A type I error is the probability of concluding that Edward has improved his ability to correctly identify types of flowers when he has not.

(d) The test statistic 4 does not lie in the critical region which is $X \geq 5$ so there is insufficient evidence to reject the null hypothesis. There is insufficient evidence to conclude that he has improved his ability to correctly identify wild flowers.

Exam practice

1 A test statistic X, has the binomial distribution $B(30, 0.2)$.
The null hypothesis for the test is $\mathbf{H}_0: p = 0.2$ and the alternative hypothesis is $\mathbf{H}_1: p < 0.2$.
If the critical region is defined as $X \leq 3$:
(a) Determine the significance level of this critical region. [1]
(b) Explain what is meant by a type II error. [1]
(c) It has been found that the actual probability is not 0.2 but is instead 0.15. Calculate the probability of a type II error. [2]

2 The NHS says it is not safe to drink 14 units or more of alcohol per week.
James has looked at some research on drinking habits of students and it said that 30% of students drink more than the safe 14 units of alcohol per week.
By modelling the number of students who drink more than 14 units of alcohol per week in a sample size of n as a binomial distribution with $B(n, 0.30)$, find:

(a) (i) The probability that 5 students in James' class of 55 students exceed the safe limit for alcohol. [1]
    (ii) The probability that at least 10 and fewer than 12 in James' class exceed the safe limit for drinking alcohol. [2]

(b) James thinks more than 30% of students drink 14 or more units of alcohol per week. He conducts his own survey and finds that out of 20 students, 8 students drink more than 14 units of alcohol per week.
Conduct a significance test at the 5% significance level and state your conclusion in context. [5]

3 Dafydd thinks that he can predict the outcome when a fair coin is tossed more often than not. In order to investigate this theory, he sets up the following hypotheses:

$H_0 : p = 0.5$ versus $H_1 : p > 0.5$

where p denotes the probability that he predicts the outcome, that is 'heads' or 'tails', correctly.

(a) He decides initially to ask a friend to toss a fair coin 20 times. Then if x denotes the number of correct predictions, he will accept H_1 if $x \geq 14$.
    (i) Find the corresponding significance level.
    (ii) Find the probability of reaching the correct conclusion if $p = 0.7$. [7]

(b) He now decides to ask a friend to toss a fair coin 50 times. He predicts the outcome correctly on 28 occasions.
    (i) Find the p-value of this result.
    (ii) Interpret your value in context. [7]

4 Bryn wants to test a coin as he thinks it is biased towards heads. He would like to perform a hypothesis test using a 5% level of significance. He tosses the coin 10 times and obtains a head 8 times.

(a) (i) Write suitable hypotheses for this test.
    (ii) State a suitable test statistic that he could use. [2]

(b) Explain what is meant by the critical region. [1]

(c) Find the critical region for this test. [2]

(d) State what conclusion can be drawn from the critical region. [1]

5 A test statistic can be modelled by the binomial distribution B(20, 0.30). The null hypothesis is $H_0 : p = 0.30$ and the alternative hypothesis is $H_1 : p < 0.30$.
Find the critical region if the significance level is 5%. [8]

6 Charlie is given a coin and he is told that it is biased so that the probability, p, of obtaining a head when tossed is 0.75. To test this, he defines the following hypotheses.
$$\mathbf{H_0} : p = 0.75; \qquad \mathbf{H_1} : p \neq 0.75$$
(a) He decides to toss the coin 50 times and he denotes the number of heads obtained by x. He defines the following critical region. $(x \leq 31) \cup (x \geq 44)$.
   (i) Determine the significance level of this test.
   (ii) Find the probability of failing to reject $\mathbf{H_0}$ if the value of p is actually 0.5. [7]
(b) In a further attempt to test whether or not the value of p is 0.75, he decides to toss the coin 200 times. He obtains 139 heads.
   (i) Calculate the approximate p-value of this result.
   (ii) Interpret the p-value. [7]

7 Gwilym buys a new computer game. He claims that he wins, on average, 60% of games played. His friend Huw believes that Gwilym wins less than 60% of games played.
(a) To investigate these conflicting claims, Gwilym plays the game 20 times and wins 7 of them.
   (i) State suitable hypotheses for testing these claims.
   (ii) Determine the p-value of the above result and state your conclusion in context. [7]
(b) During the following week, Gwilym plays the game 80 times and wins 37 of them. Determine the p-value and state your conclusion in context. [7]

8 A drug is known to cure 70% of patients suffering from a certain disease. A pharmaceutical company has developed a new drug which is claimed to cure a higher percentage than this. To test this claim, the new drug is given to 50 patients.
(a) State suitable hypotheses. [1]
(b) It is found that 40 of these patients are cured of the disease. Find the p-value of this result and state your conclusion in context. [5]
(c) The company decides to carry out a larger trial in which the new drug is to be given to 250 patients. Let x denote the number of patients cured. Given that the critical region is $x \geq 190$, find:
   (i) the significance level,
   (ii) the probability of concluding that the new drug does not increase the percentage of patients cured when in fact the percentage cured has increased to 80%. [10]

9 A fair hexagonal spinner is numbered 1 to 6. David spins it 30 times and the number of times it landed on a 6 was counted.

(a) Write suitable null and alternative hypotheses to test whether the spinner is biased towards or away from the number six. [2]

(b) A test was performed, and the spinner was spun 30 times and it landed on a six a total of 9 times. Test at the 10% level of significance whether there is evidence that the spinner is biased. [3]

(c) Find the critical region if a 10% level of significance is used. [4]

(d) Find the probability of a Type I error occurring. [2]

(e) Find the probability of a Type II error occurring if the spinner is biased in favour of six with the probability of a six occurring equal to 0.5. [2]

15 Kinematics

Prior knowledge

You will need to make sure you fully understand the following from your GCSE studies:

- Construction and interpretation of travel graphs

- Constructing tangents to curves to find velocity or acceleration

Quick revision

Displacement/distance–time graphs

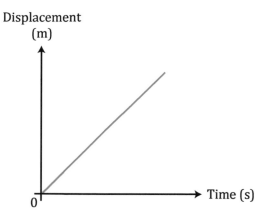

The gradient represents velocity.

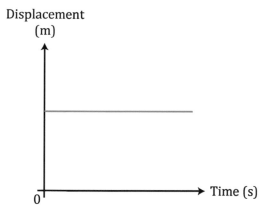

A horizontal line has zero gradient and represents a body at rest.

Velocity–time graphs

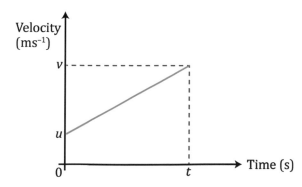

The gradient represents acceleration.

The area under the graph represents the displacement.

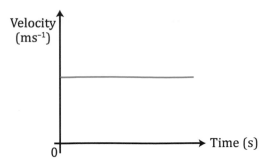

A horizontal line represents constant velocity.

Constant acceleration formulae (the *suvat* equations)

The following formulae, called the equations of motion should only be used if it is known that the motion is under constant acceleration in a straight line.

s = displacement
u = initial velocity
v = final velocity
a = acceleration
t = time

$$v = u + at$$

$$s = ut + \frac{1}{2}at^2$$

$$v^2 = u^2 + 2as$$

$$s = \frac{1}{2}(u + v)t$$

Note that all the equations of motion will need to be remembered.

Vertical motion under gravity

The acceleration due to gravity of 9.8 m s^{-2} acts vertically down.

Remember to decide on which direction to take as positive and then use the equations of motion to find unknown values.

Sketching and interpretation of velocity–time graphs

- When asked to sketch a velocity–time graph:
- Use the equations of motion to find unknown quantities if needed.
- Remember you are creating a sketch, so you do not need a scale along each axis. You only need to include the important values or values asked for in the question.
- Remember to label both sets of axes with the title of the axis and unit.
- Many of these graphs are trapezium-shaped and the displacement will be the area of a trapezium, which can be worked out using the formula:

Area of trapezium $= \frac{1}{2}\left(\text{sum of the two parallel sides}\right)$
$\times$ perpendicular distance between them.

Kinematics

The following formulae are used for motion in a straight line when the acceleration varies with time:

$$v = \frac{\mathrm{d}r}{\mathrm{d}t}$$

$$a = \frac{\mathrm{d}v}{\mathrm{d}t} = \frac{\mathrm{d}^2 r}{\mathrm{d}t^2}$$

$$r = \int v \, \mathrm{d}t$$

$$v = \int a \, \mathrm{d}t$$

These kinematics formulae will be given on the formula sheet.

Looking at exam questions

1   A vehicle moves along a straight horizontal road. Points A and B lie on the road. As the vehicle passes point A, it is moving with constant speed 15 m s^{-1}. It travels with this constant speed for 2 minutes before a constant deceleration is applied for 12 seconds so that it comes to rest at point B.

   (a)   Find the distance AB.   [3]

   The vehicle then reverses with a constant acceleration of 2 m s^{-2} for 8 seconds, followed by a constant deceleration of 1.6 m s^{-2}, coming to rest at the point C, which is between A and B.

   (b)   Calculate the time it takes for the vehicle to reverse from
          B to C.   [4]

   (c)   Sketch a velocity–time graph for the motion of the vehicle.   [3]

   (d)   Determine the distance AC.   [2]

Thinking about the question

This is a question about the motion of a vehicle with constant acceleration or zero acceleration, so the equations of motion can be used. We need to draw a velocity–time graph to model the various stages of the motion.

Starting the solution

For part (a) a velocity–time graph needs to be drawn showing the constant speed and the constant deceleration. The distance travelled will be the area under the velocity–time graph.

For part (b) we can use the equations of motion to firstly find the velocity after the period of acceleration and then use this as the initial velocity to find the time for the period of deceleration. We can add the times for the periods of acceleration and deceleration together.

For part (c) we need to add the periods of acceleration and deceleration to the graph drawn for part (a). It is important to realise that as the vehicle is reversing, the velocity needs to be negative.

For part (d) we need to find the distance whilst travelling in the opposite direction. This distance needs to be subtracted from distance AB to find the distance AC.

The solution

(a)

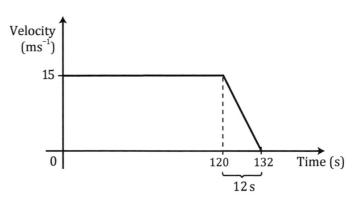

You could also have found the distances using the equations of motion rather than by drawing a graph.

Distance AB = area under the velocity–time graph

$$= 120 \times 15 + \frac{1}{2} \times 12 \times 15$$

$$= 1890 \, \text{m}$$

(b) Using $v = u + at$ with $u = 0$, $a = 2$ and $t = 8$.

$$v = 0 + 2 \times 8 = 16 \, \text{m s}^{-1}$$

Using $\quad t = \dfrac{v - u}{a} = \dfrac{0 - 16}{-1.6} = 10 \, \text{s}$

Time taken to reverse from B to C = 8 + 10 = 18 s

(c)

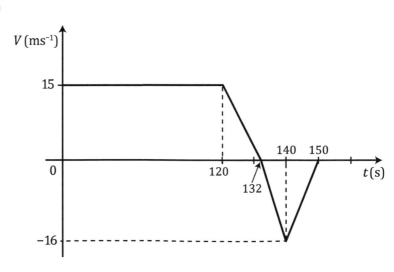

(d)   Distance AB = 1890 m

Distance BC = 0.5 × 18 × 16 = 144 m

Distance AC = 1890 − 144 = 1746 m

> Remember that the vehicle reversed direction and moved back towards A. This means the distance represented by the area under the horizontal axis must be subtracted from the distance AB to find the distance AC.

---

2   A raindrop A falls freely from rest from the top of a cliff. After it has fallen a distance 0.1 m, a second raindrop B begins to fall from rest from the top of the same cliff. The height of the cliff is 40 m.

(a)   Find the velocity of A at the instant B begins to fall. [3]

(b)   Find the velocity of A at the instant it reaches the ground. [2]

(c)   Calculate the distance between the raindrops when the first raindrop A hits the ground. [7]

Thinking about the question

This question is about motion under gravity. We need to consider which direction to take as positive. As gravity acts downwards it make sense to make the downward direction positive.

Starting the solution

For part (a) we need to use one of the equations of motion and use the distance travelled to work out the velocity.

For part (b) we need to use an equation of motion to find the velocity of the particle after travelling a distance of 40 m from rest.

For part (c) we can find the time A takes to reach the ground using our final velocity obtained from part (b). We can then find the distance B travels in this time. We can then subtract the distance B travels from the 40 m that A has travelled when it hits the ground.

The solution

(a)   Using $v^2 = u^2 + 2as$ with $u = 0$, $a = 9.8$ and $s = 0.1$.

$$v^2 = 0^2 + 2(9.8)(0.1)$$

$$v = 1.4 \text{ m s}^{-1}$$

> As we have taken the downward direction as positive, the acceleration of gravity acts in this direction so it will be positive.

(b)   Using $v^2 = u^2 + 2as$ with $u = 0$, $a = 9.8$ and $s = 40$.

$$v^2 = 0^2 + 2(9.8)(40)$$

$$v = 28 \text{ m s}^{-1}$$

(c)   Time of travel of B = time for A to reach ground.

Using $v = u + at$ with $u = 1.4$, $v = 28$ and $a = 9.8$

$$t = \frac{v - u}{a} = \frac{28 - 1.4}{9.8} = 2.7143 \text{ s}$$

> Here we are finding the time taken for A to reach the ground.

Distance travelled by B in this time is given by

$$s = ut + \tfrac{1}{2}at^2$$

$$= 0 + \tfrac{1}{2} \times 9.8 \times (2.7143)^2$$

$$= 36.1 \text{ m}$$

Distance between A and B = 40 − 36.1

$$= 3.9 \text{ (m)}$$

---

3   A particle moves along the horizontal x-axis so that its velocity v m s^{-1} at time t seconds is given by $v = 6t^2 - 8t - 5$. At time $t = 1$, the particle's displacement from the origin is −4 m. Find an expression for the displacement of the particle at time t seconds.

[3]

Thinking about the question

The velocity varies with time so the acceleration is not constant, so we have to use calculus.

Starting the solution

We need to integrate the velocity expression to give an expression for the displacement. The result will include a constant of integration. We know that the displacement is −4 m when $t = 1$ so we substitute both these into the equation to find the value of the constant of integration c. The value of c is substituted back into the displacement equation to give the final answer.

The solution

$$v = 6t^2 - 8t - 5$$

$$r = \int v\,dt$$

$$= \int (6t^2 - 8t - 5)\,dt$$

$$= \frac{6t^3}{3} - \frac{8t^2}{2} - 5t + c$$

$$= 2t^3 - 4t^2 - 5t + c$$

> We now look back at the question to see how we are going to find the value of the constant of integration, c.

When $t = 1$, $r = -4$

Hence, $-4 = 2(1)^3 - 4(1)^2 - 5(1) + c$

$$c = 3$$

$$r = 2t^3 - 4t^2 - 5t + 3$$

Exam practice ──────────────────────

1 A car is moving at a constant speed of 20 m s^{-1} when it passes the point A on a straight horizontal road. As it passes point A, it accelerates at a constant rate until its speed reaches 25 m s^{-1} in 720 m. It then decelerates at a constant rate for 4.5 minutes before stopping at the point B.
   (a) Calculate the time taken during acceleration. [3]
   (b) Sketch a velocity–time graph for the car's journey between A and B. [3]
   (c) Find the distance between A and B. [3]

2 A small steel ball bearing is dropped from rest from the top of a building 140 m tall.
   (a) Calculate the speed with which it hits the ground. [2]
   (b) Find the time taken for the ball bearing to reach the ground. [2]
   (c) State one assumption you have made in your answers. [1]

3 A car travels in a straight line and t seconds after passing a point P it has a velocity given by $v = 64 - \frac{1}{27}t^3$.
   The car comes to rest at point Q.
   (a) Show that the car comes to rest after $t = 12$ s. [3]
   (b) Calculate the distance PQ. [4]

4 A particle of mass 2 kg moves along the horizontal x-axis under the action of a resultant force F N. The particle has a velocity v m s^{-1} at time t seconds given by:
$$v = 2t^2 - 7t + 5$$
   (a) Determine the times when the particle is instantaneously at rest. [3]
   (b) The particle is at the origin when $t = 1$, find an expression for the displacement of the particle from the origin at time t s. [4]
   (c) Find an expression for the acceleration of the particle at time t s. [2]

5 A boy throws a pebble from the top of a cliff 70.2 m high with an initial velocity of 14.7 m s^{-1} vertically upwards.
   (a) Calculate the speed of the pebble 2 s after it has been thrown. [3]
   (b) Calculate the speed of the pebble when it hits the ground at the foot of the cliff. [3]
   (c) For how long is the pebble at least 3.969 m above the top of the cliff? [4]

6 A ball is dropped vertically from the top of a tower of height 24 m and, at the same instant, another ball is thrown vertically upwards from the ground so as to hit the first ball. The initial speed of this second ball is 15 m s⁻¹.
   (a) Find:
       (i) The time when the balls collide.   [5]
       (ii) The height at which they collide.   [3]
   (b) State one modelling assumption you have made in your answers.   [1]

7 The graph below shows the displacement–time graph for the motion of a particle.

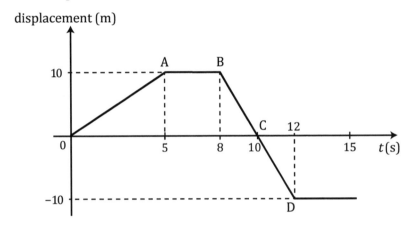

   (a) Find the velocity of the particle between O and A.   [1]
   (b) Find the velocity of the particle between B and D.   [1]
   (c) Find:
       (i) the total distance,   [3]
       (ii) the displacement in the first 12 s   [2]

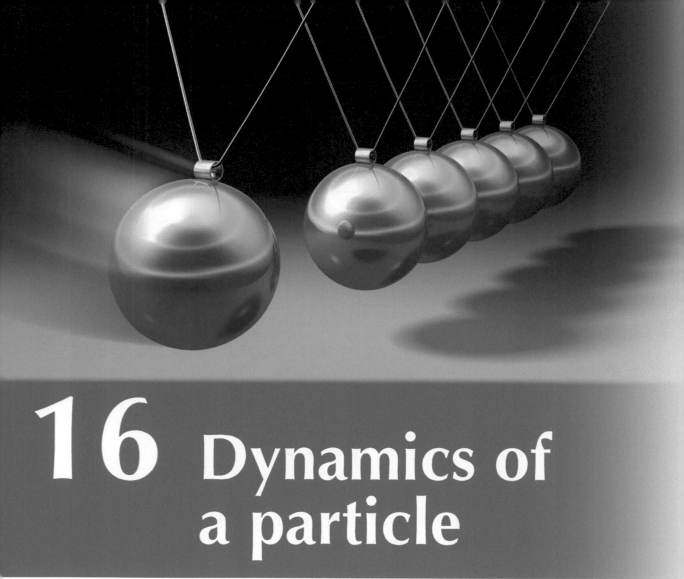

16 Dynamics of a particle

Prior knowledge

- None of this work was covered at GCSE.

Quick revision

Newton's second law of motion

Unbalanced forces produce an acceleration according to the equation

$$\text{force} = \text{mass} \times \text{acceleration} \quad \text{or for short} \quad F = ma$$

Lifts accelerating, decelerating and travelling with constant velocity

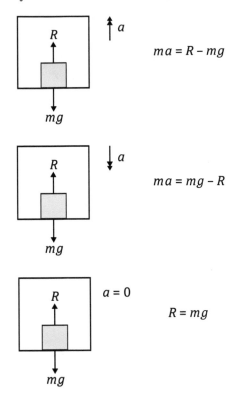

$$ma = R - mg$$

$$ma = mg - R$$

$$R = mg$$

The motion of particles connected by strings passing over fixed pulleys or pegs

Pulleys or pegs are smooth so no frictional forces act.

Strings are light and inextensible so the tension remains constant.

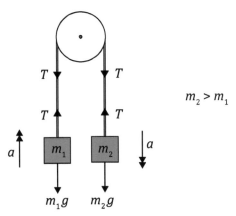

$$m_2 > m_1$$

Acceleration of each mass is the same as the string is taut.

Newton's second law can be applied to each mass separately.

For m_1, $m_1 a = T - m_1 g$

while for m_2, $m_2 a = m_2 g - T$

Looking at exam questions

1   The diagram shows two objects, A and B, of mass 3 kg and 5 kg respectively, connected by a light inextensible string passing over a light smooth pulley fixed at the end of a smooth horizontal surface. Object A lies on the horizontal surface and object B hangs freely below the pulley.

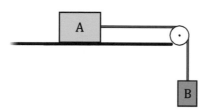

Initially, B is supported so that the objects are at rest with the string just taut. Object B is then released.

(a)   Find the magnitude of the acceleration of A and the tension in the string.   [6]

(b)   State briefly what effect a rough pulley would have on the tension in the string.   [1]

Thinking about the question

We need to first draw a labelled diagram showing the forces acting and their directions. We also need to show the direction of the acceleration which will be to the right.

Starting the solution

After marking all the forces on the diagram and the direction of the acceleration, we need to apply Newton's 2nd law of motion to each mass in turn as there are two unknowns, the acceleration and the tension in the string. The resulting equations can be solved simultaneously.

All mechanics answers make certain assumptions to simplify things. In this case, if the pulley were not smooth, the size of the tensions either side of the pulley would not be the same.

The solution

(a) Adding the forces to the diagram, we obtain:

> Redraw the diagram given and add all the forces acting. Also include the directions of the accelerations of the two masses.

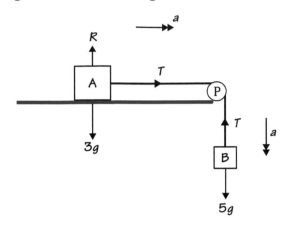

> You can treat each mass separately and then apply Newton's 2nd law to each mass. Doing this allows you to obtain two equations containing two unknowns (usually a and T) which can be solved simultaneously.

Applying Newton's 2nd law to mass A, we have

$$T = 3 \times a$$
$$= 3a$$

Applying Newton's 2nd law to mass B, we have

$$5a = 5g - T$$

Substituting $3a$ for T, we obtain: $5a = 5g - 3a$

Hence, $a = \frac{5}{8}g = 6.125 \text{ m s}^{-2}$

$$T = 3a = 3 \times 6.125 = 18.375 \text{ N}$$

(b) If the pulley is rough, the tension in the string on either side of the pulley would not be the same.

---

2 A person, of mass 68 kg, stands in a lift which is moving upwards with constant acceleration. The lift is of mass 770 kg and the tension in the lift cable is 8000 N.

(a) Determine the acceleration of the lift, giving your answer correct to two decimal places. [3]

(b) State whether the lift is getting faster, staying at the same speed or slowing down. [1]

(c) Calculate the magnitude of the reaction of the floor of the lift on the person. [3]

Thinking about the question

We need to first draw a diagram of the lift, marking on all the forces and the direction of the acceleration.

Starting the solution

For part (a) we note that the acceleration is upward, which means the resultant force on the lift will be upward. As we are just looking at the lift, the resultant force will equal the tension in the lift minus the weight of the lift and person.

For part (b) we have to look at the sign for the acceleration in part (a). If it is negative it means the lift is decelerating so the velocity is getting smaller.

For part (c) we apply Newton's 2nd law to the person in the lift.

The solution

(a)

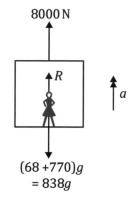

Applying Newton's 2nd law to the lift and person
$$838a = T - 838g$$

Now $\qquad T = 8000\text{ N}$

So $\qquad 838a = 8000 - 838 \times 9.8$

Hence, $\qquad a = -0.253\,...$

$\qquad\qquad = -0.25\text{ m s}^{-2}$ (2 d.p.)

(b) As we have taken upwards direction as positive, the acceleration is acting downwards so it must be a deceleration, meaning the lift is slowing down.

(c) Applying Newton's 2nd law to the person:
$$68a = R - 68g$$
$$68 \times (-0.253\,...) = R - 68 \times 9.8$$
$$R = 649\text{ N}$$

Watch out

As there is a person in the lift, their weight must be added to the weight of the lift.

The negative sign indicates a deceleration.

If you are taking the upward direction as positive, you must make sure that you include the negative sign in the acceleration.

Exam practice

1 A lift, of mass 600 kg, travels downward non-stop from the top of a building to the ground floor. It starts from rest and accelerates downwards with constant acceleration of $0.4\,\mathrm{m\,s^{-2}}$, then moves at constant speed before decelerating to rest.
   (a) Calculate the tension in the lift cable when the lift is accelerating. [3]
   (b) Find the tension in the lift cable when the lift is moving at a constant speed. [1]

2 (a) An empty lift of mass 800 kg is moving downwards with an acceleration of $0.3\,\mathrm{m\,s^{-2}}$. Calculate the tension in the lift cable. [3]
   (b) A man of mass 50 kg stands in the lift. Find the reaction of the floor on the man if the lift is moving upwards with an acceleration of $0.2\,\mathrm{m\,s^{-2}}$. [3]
   (c) Give one modelling assumption you have made in order to arrive at your answers. [1]

3 The diagram shows two bodies A and B, of mass 6 kg and 2 kg respectively, connected by a light inextensible string passing over a smooth light pulley fixed at the edge of a smooth horizontal table. Body A hangs freely below the pulley, and body B is on the table.

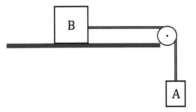

Initially, A is supported so that the system is at rest with the string taut. When A is released, it descends with uniform acceleration $a\,\mathrm{m\,s^{-2}}$. Calculate the value of a and the tension in the string. [5]

4 The mass of a lift is 5600 kg. The lift starts from rest and descends with uniform acceleration for 8 s until it reaches a speed of $V\,\mathrm{m\,s^{-1}}$. The tension in the lift cable is 50 400 N.
   (a) Show that the magnitude of the acceleration of the lift is $0.8\,\mathrm{m\,s^{-2}}$. [2]
   (b) Find the value of V. [2]

The lift maintains this constant speed of $V\,\mathrm{m\,s^{-1}}$ for 25 s before decelerating uniformly to rest. The total time for descent is 40 s.
   (c) Draw a sketch of the velocity–time graph of the motion. [3]
   (d) Calculate the total distance that the lift descends. [3]
   (e) Find the maximum tension in the lift cable during the motion. [3]

17 Vectors

Prior knowledge

You will need to make sure you fully understand Topic 9 of the AS Pure before you start this topic.

Quick revision

Scalar quantities have magnitude (i.e. size) only and include distance and speed.

Vector quantities have both magnitude and direction and include displacement, velocity, acceleration and force.

Vectors are typed in bold and not in italics, so **s**, **r**, **v**, **a** and **F** are all vectors.

The resultant of vectors acting at a point can be found by adding the individual vectors.

The magnitude of a vector

The vector $\mathbf{r} = a\mathbf{i} + b\mathbf{j}$ has magnitude given by $|\mathbf{r}| = \sqrt{a^2 + b^2}$

The direction of a vector

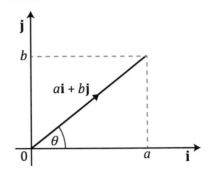

The angle made by the vector $a\mathbf{i} + b\mathbf{j}$ to the unit vector $\mathbf{i}$ is θ, where

$$\theta = \tan^{-1}\left(\frac{b}{a}\right)$$

Converting from magnitude and direction to a vector

If you know the magnitude and direction of a vector you can convert this to vector form in the following way

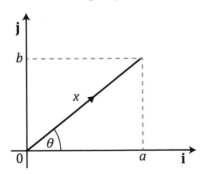

If the length of the vector is x and it is inclined at an angle θ to the $\mathbf{i}$-direction (or positive x-axis) then using trigonometry

$$\frac{a}{x} = \cos\theta°, \text{ so } a = x\cos\theta° \quad \text{and} \quad \frac{b}{x} = \sin\theta°, \text{ so } b = x\sin\theta°$$

Vector is $\mathbf{x} = a\mathbf{i} + b\mathbf{j}$

Looking at exam questions

1 Three forces **L**, **M** and **N** are given by

$$\mathbf{L} = 2\mathbf{i} + 5\mathbf{j},$$
$$\mathbf{M} = 3\mathbf{i} - 22\mathbf{j},$$
$$\mathbf{N} = 4\mathbf{i} - 23\mathbf{j}.$$

Find the magnitude and direction of the resultant of the three forces. [6]

Thinking about the question

The resultant of a series of vectors can be found by adding the vectors. We can then find the magnitude of the resultant by applying Pythagoras' theorem. Trigonometry is used to find the angle.

Starting the solution

Add the vectors by totalling up the **i** vectors and then adding this to the total for the **j** vectors.

To find the magnitude of the resultant we square the coefficient of the **i** and **j** vectors and add these together and then find the square root of the result.

To find the angle we need to draw a diagram showing the resultant vector. We can then use trigonometry to find the angle. We need to make it clear which angle we are referring to.

The solution

$$\mathbf{R} = \mathbf{L} + \mathbf{M} + \mathbf{N}$$
$$= 2\mathbf{i} + 5\mathbf{j} + 3\mathbf{i} - 22\mathbf{j} + 4\mathbf{i} - 23\mathbf{j} = 9\mathbf{i} - 40\mathbf{j}$$
$$|\mathbf{R}| = \sqrt{9^2 + (-40)^2}$$
$$= 41 \text{ N}$$

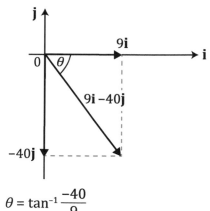

$$\theta = \tan^{-1}\frac{-40}{9}$$

$= -77.3°$ (in the direction shown on the diagram) or $282.68°$ (measured from the positive **i** axis in the anticlockwise direction).

Watch out

You need to be clear about which angle you are referring to.

Exam practice

① Three forces **P**, **Q** and **R** act on an object such that:
$$P = 3i + j$$
$$Q = 11i - 4j$$
$$R = -2i + 8j$$
The object has a mass of 5 kg. Find the magnitude of the acceleration due to the action of these three forces. [3]

② A particle having mass 2 kg moves under the action of the following forces F_1 and F_2 where
$$F_1 = (3i + 5j) \text{ N} \qquad\qquad F_2 = (5i - j) \text{ N}$$
(a) Calculate the angle the resultant force makes with the direction of **i**. [3]
(b) Find the magnitude of the acceleration of the particle, giving your answer as an exact value. [2]

③ A mass of 5 kg is subjected to the following three forces:
$$F = 4i - 2j$$
$$G = i + 7j$$
$$H = ai - bj$$
The resultant force acting on the mass is zero.
(a) Write down the exact magnitude of force **H**. [4]
(b) If forces **F** and **G** are removed, find the magnitude of the acceleration. [2]

④ A helicopter sets off from point O and is then subject to the following consecutive displacements measured in km:
$$20i + 45j \qquad \text{and} \qquad -5i + 25j$$
Find:
(a) The total displacement. [4]
(b) The total distance travelled giving your answer to 3 significant figures. [2]

⑤ The resultant force of the two forces **F** = 2**i** − 3**j** and **R** = a**i** + b**j** is 5**i** + 2**j**.
(a) Find the values of a and b. [3]
(b) Another force **Q** is added to **F** and **R** so that the resultant force now becomes zero. Give an expression for force **Q**. [2]

⑥ A particle of mass 10 kg lies at rest on a horizontal surface. It is then subjected to the two forces, **P** and **Q**, which act in the horizontal plane:
$$P = \begin{pmatrix} 3 \\ 2 \end{pmatrix} \text{ N} \qquad\qquad R = \begin{pmatrix} -1 \\ 4 \end{pmatrix} \text{ N}$$
(a) Find the magnitude of the acceleration of the particle giving your answer as an exact value. [3]
(b) Find the distance travelled by the particle in 3 s giving your answer to two decimal places. [2]

Exam practice answers

Topic 1

1 (a) If n is even, then it has 2 as a factor.
Let $n = 2x$

$$10n^2 + 5n = 10(2x)^2 + 5(2x)$$
$$= 40x^2 + 10x$$
$$= 10(4x^2 + x)$$

This expression has 10 as a factor so we have proved $10n^2 + 5n$ has a factor of 10 for any even number n.

(b) If $n = 1$, $10n^2 + 5n = 10(1)^2 + 5(1) = 15$
which does not have a factor of 10.
Hence the statement is false.

> Proofs often involve odd or even numbers. Remember that even numbers can be written as $2x$ and odd numbers can be written as $2x + 1$.

2 If you try substituting positive values in for x and y, the statement is true, so we need to try negative values.

Suppose $x = -2$ and $y = -2$, we have

$\sqrt{xy} = \sqrt{(-2)(-2)}$ and $\sqrt{4} = 2$ and $\frac{1}{2}(x + y) = \frac{1}{2}(-2 + -2) = -2$

Now 2 is not less than or equal to -2 so we have found a counter-example and the statement is incorrect.

> If a number or letter can be taken out of a bracket, then it is a factor.

3 If the first odd number is $2n + 1$, then the second and third odd numbers will be $2n + 3$ and $2n + 5$ respectively.

Sum of consecutive odd numbers $= 2n + 1 + 2n + 3 + 2n + 5 = 6n + 9$
Now $6n$ will always be even and adding an odd number (i.e. 9) to it will make it odd.
Hence, the sum of 3 consecutive odd numbers is an odd number.

4 Every integer is either a multiple of 3, one less than a multiple of 3 or one more than a multiple of 3.

A multiple of 3 can be written as $3n$ so $(3n)^3 = 27n^3 = 9(3n^3)$ which is a multiple of 9.
One less than a multiple of 3 can be written as $3n - 1$,
so $(3n - 1)^3 = 27n^3 - 27n^2 + 9n - 1 = 9(3n^3 - 3n^2 + n) - 1$
which is one less than a multiple of 9.
One more than a multiple of 3 can be written as $3n + 1$,
so $(3n + 1)^3 = 27n^3 + 27n^2 + 9n + 1 = 9(3n^3 + 3n^2 + n) + 1$
which is one more than a multiple of 9.

5 If $\theta = 60°$ and $\varphi = 300°$, then $\cos 60° = 0.5$ and $\cos 300° = 0.5$

Now $\sin 60° = \frac{\sqrt{3}}{2}$ and $\sin 300° = -\frac{\sqrt{3}}{2}$

Hence we have found a counter-example, so the statement is false.

> When substituting in the angles, don't just use acute angles.

⑥ If $a = 3$ and $c = 12$ and if $b = 5$ and $d = 15$ then $a + b = 8$ and $c + d = 27$.
Now 8 is not a factor of 27, so we have proved by counter-example that the statement is false.

Proof by deduction is used here.

⑦ $n^2 - 2n + 2 = (n - 1)^2 - 1 + 2$
$\qquad\qquad\quad = (n - 1)^2 + 1$
Now $(n - 1)^2$ is always positive for all real values of n. Adding 1 to this will keep it positive.
Hence for all real values of n, $n^2 - 2n + 2$ is positive.

Notice here that a range of values is given. As n is an integer and the range is small, we can test each value. Hence we need to use proof by exhaustion.

⑧ Using proof by exhaustion:
$\qquad n = 2, \quad n^2 + 2 = 6$
$\qquad n = 3, \quad n^2 + 2 = 11$
$\qquad n = 4, \quad n^2 + 2 = 18$
None of the above numbers are multiples of 4 so for $2 \le n \le 4$, $n^2 + 2$ is *not* a multiple of 4.

Topic 2

Here we multiply the last two brackets first.

① $(\sqrt{5} - \sqrt{3})^3 = (\sqrt{5} - \sqrt{3})(\sqrt{5} - \sqrt{3})(\sqrt{5} - \sqrt{3})$
$\qquad\qquad\quad = (\sqrt{5} - \sqrt{3})(5 - \sqrt{15} - \sqrt{15} + 3)$
$\qquad\qquad\quad = (\sqrt{5} - \sqrt{3})(8 - 2\sqrt{15})$
$\qquad\qquad\quad = 8\sqrt{5} - 2\sqrt{75} - 8\sqrt{3} + 2\sqrt{45}$
$\qquad\qquad\quad = 8\sqrt{5} - 2\sqrt{25 \times 3} - 8\sqrt{3} + 2\sqrt{9 \times 5}$
$\qquad\qquad\quad = 8\sqrt{5} - 10\sqrt{3} - 8\sqrt{3} + 6\sqrt{5}$
$\qquad\qquad\quad = 14\sqrt{5} - 18\sqrt{3}$
$\qquad\qquad\quad = 2(7\sqrt{5} - 9\sqrt{3})$

Remember to take out any square factors in the square roots.

② (a) $\dfrac{2\sqrt{5} + \sqrt{3}}{\sqrt{5} + \sqrt{3}} = \dfrac{(2\sqrt{5} + \sqrt{3})(\sqrt{5} - \sqrt{3})}{(\sqrt{5} + \sqrt{3})(\sqrt{5} - \sqrt{3})}$

Multiply the numerator and denominator by the conjugate of the denominator (i.e. $\sqrt{5} - \sqrt{3}$).

$\qquad\qquad = \dfrac{10 - 2\sqrt{15} + \sqrt{15} - 3}{5 - 3}$

$\qquad\qquad = \dfrac{7 - \sqrt{15}}{2}$

$\qquad\qquad = \dfrac{1}{2}(7 - \sqrt{15})$

Note that $(x^a)^b = x^{a \times b}$

(b) $(3^a)^3 \times 3^a \times 3 = 81$
$\qquad 3^{3a} \times 3^a \times 3^1 = 3^4$
$\qquad\qquad\quad 3^{4a + 1} = 3^4$

The indices can now be equated.

$\qquad\qquad\quad 4a + 1 = 4$
$\qquad\qquad\qquad\quad a = \dfrac{3}{4}$

3 (a) $\dfrac{10}{7 + 2\sqrt{11}} = \dfrac{10}{(7 + 2\sqrt{11})} \times \dfrac{(7 - 2\sqrt{11})}{(7 - 2\sqrt{11})}$

$= \dfrac{70 - 20\sqrt{11}}{49 - 44}$

$= \dfrac{70 - 20\sqrt{11}}{5}$

$= 14 - 4\sqrt{11}$

> Multiply the top and bottom of the original fraction by the conjugate of the bottom (i.e. $7 - 2\sqrt{11}$).

> Make sure you spot that both terms in the numerator can be divided by the denominator (i.e. 5).

(b) $(4\sqrt{3})^2 - (\sqrt{8} \times \sqrt{50}) - \dfrac{5\sqrt{63}}{\sqrt{7}} = 48 - (2\sqrt{2} \times 5\sqrt{2}) - \dfrac{5\sqrt{7 \times 9}}{\sqrt{7}}$

$= 48 - 20 - 15$

$= 13$

4 As -3, 2.5 and 4 are the roots of the equation $2x^3 + ax^2 + bx + c = 0$, the factors must be $(x + 3)(2x - 5)(x - 4)$
Hence, the graph of $f(x)$ can be written as
$\qquad f(x) = (x + 3)(2x - 5)(x - 4)$
The graph of $f(x)$ intersects the y-axis at the point where $x = 0$.
Hence, when $x = 0$, we obtain
$\qquad f(x) = (0 + 3)(0 - 5)(0 - 4) = 60$
The graph intersects the y-axis at $(0, 60)$

> Note that the root $x = -2.5$ would give $x + 2.5 = 0$ which can be multiplied by 2 to give $2x - 5 = 0$.

> This means $2x - 5$ is the factor.

> If you look at $2x^3 + ax^2 + bx + c = 0$, there must be a 2 as a coefficient of x in one of the roots to give the 2 as the coefficient of x^3.

5 For real distinct roots, $b^2 - 4ac > 0$
Hence $\qquad\qquad\qquad 144 - 16m > 0$
$\qquad\qquad\qquad\qquad -16m > -144$
$\qquad\qquad\qquad\qquad\qquad m < 9$
For the equation $\quad 3x^2 + mx + 7 = 0$,
$\qquad\qquad\qquad b^2 - 4ac = m^2 - 4(3)(7)$
$\qquad\qquad\qquad\qquad\qquad = m^2 - 84$
Now as $m < 9$, $m^2 < 81$ and this means that $m^2 - 84$ would be negative.
Hence $b^2 - 4ac < 0$, so there are no real roots.

> Remember if you multiply or divide both sides of an inequality by a negative number, the inequality sign must be reversed.

6 (a) $y = f(x + 3)$ represents a translation of the curve $y = f(x)$ by $\begin{pmatrix} -3 \\ 0 \end{pmatrix}$.

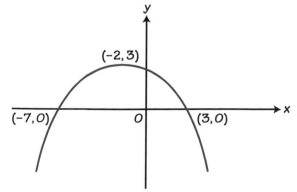

> The x-coordinates of the points marked on the curve will all shift 3 units to the left. The y-coordinates will be unaltered.

(b) $y = f(ax)$ represents a one-way stretch with scale factor $\frac{1}{a}$ parallel to the x-axis.
This stretch alters the coordinates of the points of intersection of the curve with the x-axis.
For the point $(6, 0)$ to be stretched to $(2, 0)$ the scale factor would be $\frac{1}{3}$ so $a = 3$.

For the point $(-4, 0)$ to be stretched to $(2, 0)$ the scale factor would be $\frac{1}{-2}$ so $a = -2$.

7 (a) Let $f(x) = 2x^3 + 3x^2 - 3x - 2$
Try $x = 1, f(1) = 2(1)^3 + 3(1)^2 - 3(1) - 2 = 0$ so $x - 1$ is a factor.
So, $(x - 1)(ax^2 + bx + c) = 2x^3 + 3x^2 - 3x - 2$
Equating coefficients of x^3, gives $a = 2$.
Equating coefficients independent of x, gives $c = 2$.
Equating coefficients of x^2, gives $b - a = 3$ so $b = 5$.
Hence we have $(x - 1)(2x^2 + 5x + 2)$
Factorising the quadratic, we obtain
$$(x - 1)(2x + 1)(x + 2)$$

> You may be able to spot that $a = 2$ without equating coefficients.

(b) $f(x) = 2x^3 + 3x^2 - 3x - 2$
As $2x - 1$ is a factor of $f(x)$, $2x - 1 = 0$, so $x = \frac{1}{2}$.
$$f\left(\tfrac{1}{2}\right) = 2\left(\tfrac{1}{2}\right)^3 + 3\left(\tfrac{1}{2}\right)^2 - 3\left(\tfrac{1}{2}\right) - 2 = -2\tfrac{1}{2}$$
Remainder $= -2\frac{1}{2}$

8
$$\begin{aligned}3x^2 - 12x + 15 &= 3[x^2 - 4x + 5] \\ &= 3[(x - 2)^2 - 4 + 5] \\ &= 3[(x - 2)^2 + 1] \\ &= 3(x - 2)^2 + 3\end{aligned}$$
The minimum value of this expression is at $(2, 3)$.
Hence the minimum value of $3x^2 - 12x + 15$ is $+3$.

$\dfrac{1}{3x^2 - 12x + 15}$ has its greatest value when the denominator has its smallest value.
Hence, greatest value is $\frac{1}{3}$.

9 (a) $4x^2 + 40x - 69 = 4\left[x^2 + 10x - \frac{69}{4}\right]$
$$= 4\left[(x + 5)^2 - 25 - \frac{69}{4}\right]$$
$$= 4\left[(x + 5)^2 - \frac{169}{4}\right]$$
$$= 4(x + 5)^2 - 169$$
Hence, $a = 4, b = 5, c = -169$

> Remember to specifically state the values of a, b and c.

(b) $4(x + 5)^2 - 169 = 0$

$$4(x + 5)^2 = 169$$
$$2(x + 5) = \pm 13$$
$$x + 5 = \frac{\pm 13}{2}$$
$$x = 6.5 - 5 \text{ or } -6.5 - 5$$
$$x = 1.5 \text{ or } -11.5$$

> Square root both sides of the equation remembering to include the ±.

10 $\sqrt{500} + \left(\sqrt{12} \times \sqrt{15}\right) - \dfrac{7\sqrt{60}}{\sqrt{3}} = \sqrt{100 \times 5} + \sqrt{180} - \left(\dfrac{7\sqrt{60}}{\sqrt{3}} \times \dfrac{\sqrt{3}}{\sqrt{3}}\right)$

$$= 10\sqrt{5} + \sqrt{36 \times 5} - \frac{7\sqrt{180}}{3}$$

$$= 10\sqrt{5} + 6\sqrt{5} - \frac{7\sqrt{36 \times 5}}{3}$$

$$= 10\sqrt{5} + 6\sqrt{5} - 14\sqrt{5}$$

$$= 2\sqrt{5}$$

Topic 3

1   (a) $x^2 + y^2 - 8x + 4y + 11 = 0$
$$(x - 4)^2 + (y + 2)^2 - 16 - 4 + 11 = 0$$
$$(x - 4)^2 + (y + 2)^2 - 9 = 0$$
$$(x - 4)^2 + (y + 2)^2 = 9$$
Centre is at $(4, -2)$ and radius $= 3$.

> The method using completing the squares has been used here but you could use the formula but it is tricky to remember.

(b) If the circle touches C externally, then the distance between the centres of the circles must equal the sum of their radii.
Now $x^2 + y^2 = a^2$ has centre $(0, 0)$ and radius $= a$.

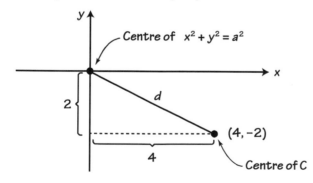

Using Pythagoras' theorem,
$$d^2 = 2^2 + 4^2$$
$$= 4 + 16$$
$$d = \sqrt{20}$$
The sum of the two radii must equal $\sqrt{20}$, so
$$a + 3 = \sqrt{20}$$
$$a = 1.47 \text{ (2 d.p.)}$$

Exam practice answers

2 (a) $x^2 + y^2 - 8x + 6y - 24 = 0$
$$(x - 4)^2 + (y + 3)^2 - 16 - 9 - 24 = 0$$
$$(x - 4)^2 + (y + 3)^2 = 49$$
Centre of C is $(4, -3)$

(b) Distance between two points $= \sqrt{(x_2 - x_1)^2 + (y_2 - y_1)^2}$

Remember this formula.

Distance between $(4, -3)$ and $(-4, 3)$ $= \sqrt{(-4 - 4)^2 + (3 - -3)^2}$
$$= \sqrt{64 + 36}$$
$$= 10$$

As this distance is greater than the radius of the circle (i.e. 7), point P lies outside the circle.

3 (a) For C_1 $\qquad x^2 + y^2 - 2x + 8y - 8 = 0$
$$(x - 1)^2 + (y + 4)^2 - 1 - 16 - 8 = 0$$
$$(x - 1)^2 + (y + 4)^2 = 25$$
Centre of C_1 is $(1, -4)$

For C_2 $\qquad x^2 + y^2 - 2x + 8y - 19 = 0$
$$(x - 1)^2 + (y + 4)^2 - 1 - 16 - 19 = 0$$
$$(x - 1)^2 + (y + 4)^2 = 36$$
Centre of C_2 is $(1, -4)$
Hence, both circles share the same centre.

(b) (i) If point P lies on C_1, its coordinates will satisfy the equation of the circle.
$$x^2 + y^2 - 2x + 8y - 8 = 16 + 0 - 8 + 0 - 8 = 0$$
P lies on the circle.

As the radius to point P and the tangent at P are perpendicular, the product of their gradients will equal -1.

(ii) Gradient of radius to point $P = \dfrac{y_2 - y_1}{x_2 - x_1} = \dfrac{-4 - 0}{1 - 4} = \dfrac{4}{3}$

Gradient of tangent at $P = -\dfrac{3}{4}$

Equation of tangent at P:
$$y - y_1 = m(x - x_1)$$
$$y - 0 = -\tfrac{3}{4}(x - 4)$$
$$4y = -3x + 12$$
$$3x + 4y - 12 = 0$$

4 (a) Gradient of AB $= \dfrac{y_2 - y_1}{x_2 - x_1} = \dfrac{1 - -3}{6 - -2} = \dfrac{1}{2}$

(b) Gradient of BC $= \dfrac{y_2 - y_1}{x_2 - x_1} = \dfrac{3 - 1}{k - 6} = \dfrac{2}{k - 6}$

As AB and BC are perpendicular the product of their gradients will equal -1.
$$\left(\tfrac{1}{2}\right)\left(\dfrac{2}{k - 6}\right) = -1$$
$$1 = -k + 6$$
$$k = 5$$

(c) Gradient of line L $= \dfrac{2}{5-6} = -2$

Equation of L:
$$y - y_1 = m(x - x_1)$$
$$y - -3 = -2(x - -2)$$
$$y + 3 = -2x - 4$$
$$2x + y + 7 = 0$$

(d) When $x = 0$, $y + 7 = 0$ giving $y = -7$

Point D is $(0, -7)$ and C is $(5, 3)$

$$CD = \sqrt{(5 - 0)^2 + (3 - -7)^2}$$
$$= \sqrt{125} \text{ or } 5\sqrt{5}$$

The formula for the distance between two points is used here.

⑤ Centre of circle will be at the mid-point of AB.

Coordinates of mid-point of AB are

$$\left(\dfrac{x_1 + x_2}{2}, \dfrac{y_1 + y_2}{2}\right) = \left(\dfrac{2 + 0}{2}, \dfrac{1 + -5}{2}\right) = (1, -2)$$

Centre of circle is $(1, -2)$

Distance between two points $= \sqrt{(x_2 - x_1)^2 + (y_2 - y_1)^2}$

Distance between $(1, -2)$ and $(2, 1)$
$$= \sqrt{(2 - 1)^2 + (1 - -2)^2}$$
$$= \sqrt{1 + 9}$$
$$= \sqrt{10}$$

Note that you have to draw a diagram in order to see that the quadrilateral is a kite.

Equation of circle, C is $(x - 1)^2 + (y - -2)^2 = \left(\sqrt{10}\right)^2$

$$(x - 1)^2 + (y + 2)^2 = 10$$

⑥ (a) Length of line from centre of circle to origin

Distance between $(-4, 3)$ and $(0, 0) = \sqrt{(0 - -4)^2 + (0 - 3)^2}$
$$= \sqrt{25} = 5$$

Now this distance is the radius of the circle.

Equation of the circle is $(x + 4)^2 + (y - 3)^2 = 25$

At point B, $y = 0$ so substituting $y = 0$ into the above equation, we obtain:

$$(x + 4)^2 + 9 = 25$$
$$(x + 4)^2 = 16$$
$$x + 4 = \pm 4$$

Giving $x = 0$ or -8

Hence B is $(-8, 0)$

Gradient of the line joining B to the centre of the circle

$$= \dfrac{3 - 0}{-4 - -8} = \dfrac{3}{4}$$

Gradient of tangent $= -\dfrac{4}{3}$

$$\text{Equation of tangent is} \quad y - 0 = -\tfrac{4}{3}\left(x - -8\right)$$
$$y = -\tfrac{4}{3}\left(x + 8\right)$$
$$3y = -4x - 32$$
$$4x + 3y + 32 = 0$$

(b) At the point where tangent intersects the y-axis, $x = 0$.
$$4(0) + 3y + 32 = 0$$
$$y = -\tfrac{32}{3}$$

Coordinates are $\left(0, -\tfrac{32}{3}\right)$

7 (a)

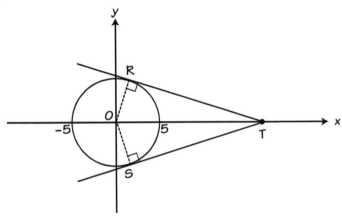

Quadrilateral ORTS is a kite.

Note that you have to draw a diagram in order to see that the quadrilateral is a kite.

(b) OR = OS = 5 as they are both radii
OT = 11
$OT^2 = OR^2 + RT^2$
$11^2 = 5^2 + RT^2$
$RT = 4\sqrt{6}$

Triangles OST and ORT are congruent.

Area of triangle ORT $= \tfrac{1}{2} \times$ base $\times$ perpendicular height
$$= \tfrac{1}{2} \times 4\sqrt{6} \times 5$$
$$= 10\sqrt{6}$$

Note that you do not work out the area as a decimal as the question asks for the exact value to be found.

Both triangles forming the kite are the same.
Area of ORTS $= 2 \times 10\sqrt{6}$
$$= 20\sqrt{6}$$

Topic 4

1 (a) Looking up the formula for the expansion of $(a + b)^n$ in the formula booklet, we obtain:

$$(a + b)^n = a^n + \binom{n}{1}a^{n-1}b + \binom{n}{2}a^{n-2}b^2 + \dots$$

$$(a + b)^4 = a^4 + 4a^3b + 6a^2b^2 + 4ab^3 + b^4$$

(b) As $b = 2x$,

Term in x^2 is $6a^2(2x)^2$, so coefficient in x^2 is $24a^2$

Term in x^3 is $4a(2x)^3$, so coefficient in x^3 is $32a$

Now

$$24a^2 = 12 \times 32a$$
$$2a^2 = 32a$$
$$2a^2 - 32a = 0$$
$$a(2a - 32) = 0$$
$$a = 0 \text{ or } 16$$

If $a = 0$, there would be no expansion, so $a = 16$.

2 (a) The formula for the expansion of $(1 + x)^n$ is obtained from the formula booklet.

$$(1 + x)^n = 1 + nx + \frac{n(n - 1)}{2!}x^2 + \frac{n(n - 1)(n - 2)}{3!}x^3 + \dots$$

Putting $n = 6$ into this formula gives:

$$(1 + x)^6 = 1 + 6x + \frac{6(5)}{2!}x^2 + \frac{6(5)(4)}{3!}x^3 + \dots$$

Note that using the first three terms only provides an approximate value.

Hence,

$$(1 + x)^6 \approx 1 + 6x + \frac{6(5)}{2!}x^2 + \frac{6(5)(4)}{3!}x^3 + \dots$$

$$\approx 1 + 6x + 15x^2 + 20x^3$$

(b) $1 - 0.01 = 0.99$

So, $\quad 0.99^6 = (1 - 0.01)^6$

Putting $x = -0.01$ into the expansion of $(1 + x)^6$ gives

$$(1 - 0.01)^6 \approx 1 + 6(-0.01) + 15(-0.01)^2 + 20(-0.01)^3$$
$$\approx 0.94148$$
$$\approx 0.9415 \text{ (4 decimal places)}$$

3 The formula is as follows:

$$(a + b)^n = a^n + \binom{n}{1}a^{n-1}b + \binom{n}{2}a^{n-2}b^2 + \dots + \binom{n}{r}a^{n-r}b^r + \dots + b^n$$

Here $n = 6$, $a = x$ and $b = \dfrac{3}{x}$.

If you can remember Pascal's triangle, then finding the coefficients is easy. As n is 4, you look for the line in the triangle starting with 1 and then 4.

As the coefficient of the term in x^2 is twelve times the coefficient of the term in x^3.

Watch out

When obtaining a numerical answer, always check to see if the question asks that the answer needs to be given to a certain number of decimal places or significant figures. Marks can be lost needlessly by not doing this.

Looking at the above it can be seen that the term in x^2 is the third term in the expansion.

Exam practice answers

Substituting in the values for a, b and n we obtain:

$$\left(x + \frac{3}{x}\right)^6 = x^6 + \binom{6}{1}x^5\left(\frac{3}{x}\right) + \binom{6}{2}x^4\left(\frac{3}{x}\right)^2 + \binom{6}{3}x^3\left(\frac{3}{x}\right)^3 + \dots$$

Term in $x^2 = \binom{6}{2}x^4\left(\frac{3}{x}\right)^2$

The last line of Pascal's triangle shows the line we need as we need the second number in the line to be a 6 which is the power to which the bracket is to be raised.

To find the coefficients we will expand Pascal's triangle.

$$1$$
$$1 \quad 1$$
$$1 \quad 2 \quad 1$$
$$1 \quad 3 \quad 3 \quad 1$$
$$1 \quad 4 \quad 6 \quad 4 \quad 1$$
$$1 \quad 5 \quad 10 \quad 10 \quad 5 \quad 1$$
$$1 \quad 6 \quad 15 \quad 20 \quad 15 \quad 6 \quad 1$$

As $\binom{6}{2} = 15$, we have term in $x^2 = 15x^4\left(\frac{3}{x}\right)^2 = 135x^2$

4 (a) $(a + b)^n = a^n + \binom{n}{1}a^{n-1}b + \binom{n}{2}a^{n-2}b^2 + \binom{n}{3}a^{n-3}b^3 + \binom{n}{4}a^{n-4}b^4$

$(2x + 3)^4 = (2x)^4 + 4(2x)^3(3) + 6(2x)^2(3)^2 + 4(2x)(3)^3 + (2x)^0(3)^4$
$\qquad\qquad = 16x^4 + 96x^3 + 216x^2 + 216x + 81$

(b) $(1 + x)^n = 1 + nx + \dfrac{n(n-1)}{2!}x^2 + \dots$

For $(1 + 3x)^n$ the coefficient of x^2 is $\dfrac{n(n-1)(3x)^2}{2!}$

Now this coefficient is 54, hence $\dfrac{n(n-1) \times 9}{2!} = 54$

$$9n^2 - 9n = 108$$
$$9n^2 - 9n - 108 = 0$$
$$n^2 - n - 12 = 0$$
$$(n - 4)(n + 3) = 0$$

Note that the question says $n > 0$, so we ignore the solution $n = -3$.

Now $n > 0$, so $n = 4$.

5 $(1 + x)^4 = 1 + nx + \dfrac{n(n-1)}{2!}x^2 + \dfrac{n(n-1)(n-2)}{3!}x^3$
$$+ \dfrac{n(n-1)(n-2)(n-3)}{4!}x^4$$

$\left(1 + \sqrt{x}\right)^4 = 1 + 4\sqrt{x} + \dfrac{4(3)}{2}\left(\sqrt{x}\right)^2 + \dfrac{4(3)(2)}{3 \times 2}\left(\sqrt{x}\right)^3 + \dfrac{4(3)(2)(1)}{4 \times 3 \times 2}\left(\sqrt{x}\right)^4$

$\qquad = 1 + 4\sqrt{x} + 6x + 4x\sqrt{x} + x^2$

Topic 5

1 (a)

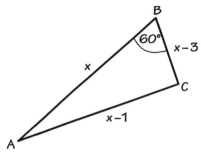

Note that AB is the longest side as it does not have a value subtracted from it such as $x - 1$.

Using the Cosine rule

$$(x - 1)^2 = x^2 + (x - 3)^2 - 2(x)(x - 3)\cos 60°$$
$$x^2 - 2x + 1 = x^2 + x^2 - 6x + 9 - x(x - 3)$$
$$x^2 - 2x + 1 = x^2 + x^2 - 6x + 9 - x^2 + 3x$$
$$x = 8$$

(b) Area of triangle $= \frac{1}{2} ab \sin C$

$$= \frac{1}{2} \times 8 \times 5 \sin 60°$$
$$= 20 \times \frac{\sqrt{3}}{2}$$
$$= 10\sqrt{3} \text{ cm}^2$$

2 $\quad 1 - \tan 2x = 3$
$$\tan 2x = -2$$
$$2x = \tan^{-1}(-2)$$

Rearranging the equation for $\tan 2x$.

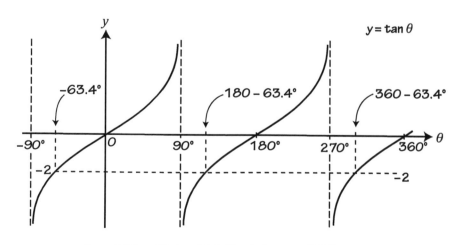

$$2x = 180 - 63.4 = 116.6°, \text{ giving } x = 58.3°$$
Or $\quad 2x = 360 - 63.4 = 296.6°, \text{ giving } x = 148.3°$

3 (a)

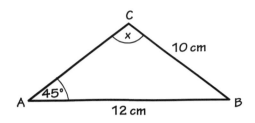

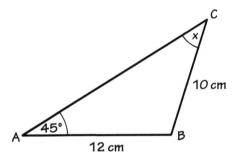

Using the Sine rule

$$\frac{\sin x}{12} = \frac{\sin 45°}{10}$$

$$\sin x = \frac{12 \sin 45°}{10}$$

$$= 0.8485$$

$$x = 58.1° \text{ or } 121.9°$$

Hence $B\hat{C}A = 58.1°$ or $121.9°$

(b) $A\hat{B}C = 180 - (45 + 58.1) = 76.9°$
or $A\hat{B}C = 180 - (45 + 121.9) = 13.1°$

Using $A\hat{B}C = 76.9°$ and using the Cosine rule, we obtain:

$AC^2 = 12^2 + 10^2 - 2 \times 12 \times 10 \cos 76.9°$

$AC = 13.8 \text{ cm}$

Using $A\hat{B}C = 13.1°$ and using the Cosine rule, we obtain:

$AC^2 = 12^2 + 10^2 - 2 \times 12 \times 10 \cos 13°$

$AC = 3.2 \text{ cm}$

4   The area is given so we use the formula Area $= \frac{1}{2} ab \sin C$

Area of triangle $= \frac{1}{2} \times 5 \times 8 \sin B\hat{A}C$

$12 = \frac{1}{2} \times 5 \times 8 \sin B\hat{A}C$

$\sin B\hat{A}C = 0.6$

$B\hat{A}C = \sin^{-1} (0.6)$

$= 36.869...$

Using the Cosine rule:

$BC^2 = 5^2 + 8^2 - 2 \times 5 \times 8 \cos 36.869...$

$BC = 5.0 \text{ cm} \text{ (1 d.p.)}$

5

$$3 \sin x = \tan x$$
$$3 \sin x = \frac{\sin x}{\cos x}$$
$$3 \sin x \cos x = \sin x$$
$$3 \sin x \cos x - \sin x = 0$$
$$\sin x (3 \cos x - 1) = 0$$
$$\sin x = 0 \quad \text{or} \quad \cos x = \frac{1}{3}$$

Don't be tempted to divide each side by $\sin x$. If you do this you will lose some of the solutions.

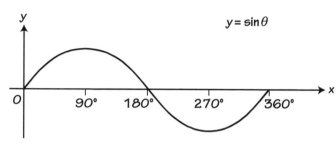

If $\sin x = 0$, $x = 0°, 180°, 360°$

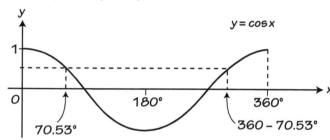

If $\cos x = \frac{1}{3}$, $x = \cos^{-1}\left(\frac{1}{3}\right)$, $x = 70.53°$ or $360 - 70.53 = 289.47°$

Hence $x = 0°, 70.53°, 180°, 289.47°, 360°$

6

$$6(1 - \cos^2 x) = 4 + \cos x$$
$$6 - 6 \cos^2 x = 4 + \cos x$$
$$6 \cos^2 x + \cos x - 2 = 0$$
$$(3 \cos x + 2)(2 \cos x - 1) = 0$$
$$\cos x = -\frac{2}{3} \quad \text{or} \quad \cos x = \frac{1}{2}$$

When $\cos x = \frac{1}{2}$, $x = 60°$ or $300°$

When $\cos x = -\frac{2}{3}$, $x = 131.81°$ or $228.19°$

Hence, $x = 60°, 131.81°, 228.19°, 300°$

7

$$3 \tan \theta + 2 \sin \theta = 0$$
$$3 \frac{\sin \theta}{\cos \theta} + 2 \sin \theta = 0$$
$$3 \sin \theta + 2 \sin \theta \cos \theta = 0$$
$$\sin \theta (3 + 2 \cos \theta) = 0$$
$$\sin \theta = 0 \quad \text{or} \quad 3 + 2 \cos \theta = 0$$

Multiply both sides by $\cos \theta$.

The smallest value $\cos \theta$ can take is -1.

Exam practice answers

The second solution gives $\cos \theta = -\frac{3}{2}$, which is impossible, so this solution is ignored.

$\sin \theta = 0$, $\theta = 0°$, $180°$ or $360°$

Topic 6

This is one of the three log laws you must be able to prove.

① (a) Suppose $x = a^n$ and $y = a^m$

Taking logs of both sides we have:

$\log_a x = n$ and $\log_a y = m$

$xy = a^n \times a^m$

$xy = a^{n+m}$

Taking logs of both sides gives:

$\log_a (xy) = n + m$

But $n = \log_a x$ and $m = \log_a y$

$\log_a (xy) = \log_a x + \log_a y$

(b) $\log_a 36 + \frac{1}{2} \log_a 256 - 2 \log_a 48 = \log_a 36 + \log_a \sqrt{256} - \log_a 48^2$

$= \log_a 36 + \log_a 16 - \log_a 2304$

$= \log_a \left(\frac{36 \times 16}{2304} \right)$

$= \log_a \left(\frac{1}{4} \right)$

(c) $2^{x+1} = 5$

Taking ln of both sides, we obtain:

$\ln 2^{x+1} = \ln 5$

$(x + 1) \ln 2 = \ln 5$

$x + 1 = \dfrac{\ln 5}{\ln 2}$

$x = \dfrac{\ln 5}{\ln 2} - 1$

$= 1.322$ (3 d.p.)

Common error – don't write $\dfrac{\ln 5}{\ln 2}$ as $\ln \dfrac{5}{2}$.

② (a) Suppose $x = (a^p)$, then $\log_a x = p$

Raising both sides of the equation to the power n gives:

$x^n = (a^p)^n$

$x^n = a^{np}$

Taking logs of both sides we have:

$\log_a x^n = np$

Hence

$\log_a x^n = n \log_a x$

As $p = \log_a x$ this can be substituted in for p.

(b) $\log_a(3x + 4) - \log_a x = 3\log_a 2$

$\log_a(3x + 4) - \log_a x = \log_a 2^3$

$\log_a(3x + 4) - \log_a x = \log_a 8$

$$\log_a\left(\frac{3x + 4}{x}\right) = \log_a 8$$

$$\frac{3x + 4}{x} = 8$$

$$3x + 4 = 8x$$

$$x = 0.8$$

(c) $4^{3y + 2} = 7$

Taking ln of both sides, we obtain:

$$\ln 4^{3y + 2} = \ln 7$$

$$(3y + 2)\ln 4 = \ln 7$$

$$3y + 2 = \frac{\ln 7}{\ln 4}$$

$$3y + 2 = 1.4037$$

$$y = -0.199 \text{ (3 d.p.)}$$

3   Let $y = 3^x$

$$9^x - 6(3^x) + 8 = 0$$

$$(3^2)^x - 6(3^x) + 8 = 0$$

$$(3^x)^2 - 6(3^x) + 8 = 0$$

Substituting $y = 3^x$, we obtain $y^2 - 6y + 8 = 0$

$$(y - 4)(y - 2) = 0$$

$y = 2$ or 4

Hence, $3^x = 2$ or $3^x = 4$

If $3^x = 2$ then $\ln 3^x = \ln 2$ so $x\ln 3 = \ln 2$, $x = \dfrac{\ln 2}{\ln 3} = 0.631$ (3 s.f.)

If $3^x = 4$ then $\ln 3^x = \ln 4$ so $x\ln 3 = \ln 4$, $x = \dfrac{\ln 4}{\ln 3} = 1.26$ (3 s.f.)

4   Suppose $x = a^n$ and $y = a^m$.

Taking logs of both sides, we obtain:

$$\log_a x = n \text{ and } \log_a y = m$$

Now

$$\frac{x}{y} = \frac{a^n}{a^m} = a^{n - m}$$

Taking logs of both sides, we obtain

$$\log_a\left(\frac{x}{y}\right) = \log_a(a^{n - m})$$

$$= n - m$$

But $m = \log_a y$ and $n = \log_a x$

Hence

$$\log_a\left(\frac{x}{y}\right) = \log_a x - \log_a y$$

The aim here is to get a single log function on either side of the equals sign. To achieve this, we need to use the rules of logs.

Taking antilogs of both sides of the equation to remove the log function.

Don't make the mistake of writing $\dfrac{\ln 7}{\ln 4}$ as $\ln\dfrac{7}{4}$.

Use the substitution $y = 3^x$ here and then form a quadratic equation.

Exam practice answers

⑤ (a) $\log_{10} 2 + 2\log_{10} 18 - \frac{3}{2}\log_{10} 36 = \log_{10} 2 + \log_{10} 18^2 - \log_{10} 36^{\frac{3}{2}}$

$$= \log_{10} 2 + \log_{10} 324 - \log_{10} 216$$

$$= \log_{10}\left(\frac{2 \times 324}{216}\right)$$

$$= \log_{10} 3$$

In order to remove the logs it is necessary to combine the logs on the right-hand side.

(b) $\log_{10}(x^2 + 48) = \log_{10} x + 2\log_{10} 4$

$\log_{10}(x^2 + 48) = \log_{10} x + \log_{10} 16$

$\log_{10}(x^2 + 48) = \log_{10} 16x$

Taking antilogs of both sides, we obtain:

$$x^2 + 48 = 16x$$
$$x^2 - 16x + 48 = 0$$
$$(x - 12)(x - 4) = 0$$
$$x = 4 \text{ or } 12$$

⑥ (a) $V = ab^t$

Taking logs to base 10 of both sides, we obtain:

$\log_{10} V = \log_{10} a + \log_{10} b^t$

$\log_{10} V = \log_{10} a + t\log_{10} b$

$\log_{10} V = t\log_{10} b + \log_{10} a$

This equation is now in the form $y = mx + c$ which is the equation of a straight line. Notice that as the variable on the x-axis is t, $\log_{10} b$ will be the gradient of the line.

(b) Gradient $= \log_{10} b = 0.02$

Solving gives $b = 10^{0.02} = 1.0471$ (correct to 4 d.p.)

Intercept $= \log_{10} a = 5.6$

Solving gives $a = 10^{5.6} = 398\,107.1706$ (correct to 4 d.p.)

(c) The gradient gives the rate of increase in price (or the price increase per year).

The intercept on the y-axis gives the initial price of the house (i.e. when it was new).

(d) $V = ab^t$

$ = 398\,107.1706 \times 1.0471^3$

$ = £457\,051$

(e) The model predicts continuous house price growth. Prices can go down as well as up, and there are many factors that can cause the rate of increase to fluctuate.

Topic 7

① (a) Let $y = (2x - 1)(x + 3)(x + 2)$

$ = (2x - 1)(x^2 + 5x + 6)$

$ = 2x^3 + 10x^2 + 12x - x^2 - 5x - 6$

$ = 2x^3 + 9x^2 + 7x - 6$

$\dfrac{dy}{dx} = 6x^2 + 18x + 7$

Here we have multiplied the last two brackets first.

Remember that $\sqrt{x} = x^{\frac{1}{2}}$.

(b) Let $y = \sqrt{x}\left(\sqrt{x} - x^2\right)$

$\qquad = x - x^{\frac{5}{2}}$

$\qquad \dfrac{dy}{dx} = 1 - \dfrac{5}{2}x^{\frac{3}{2}}$

2 $y = 6\sqrt{x} + \dfrac{25}{x^2} + 3x^5$

$\qquad\qquad = 6x^{\frac{1}{2}} + 25x^{-2} + 3x^5$

$\qquad\qquad \dfrac{dy}{dx} = 3x^{-\frac{1}{2}} - 50x^{-3} + 15x^4$

$\qquad\qquad = \dfrac{3}{\sqrt{x}} - \dfrac{50}{x^3} + 15x^4$

When $x = 1$, $\qquad \dfrac{dy}{dx} = \dfrac{3}{\sqrt{1}} - \dfrac{50}{1^3} + 15(1)^4$

$\qquad\qquad = 3 - 50 + 15$

$\qquad\qquad = -32$

3 (a) $y = (2x - 1)^2$

$\qquad y = 4x^2 - 4x + 1$

Increasing x by a small amount δx will result in y increasing by a small amount δy.

Putting $x + \delta x$ and $y + \delta y$ into the equation we have:

$\qquad y + \delta y = 4(x + \delta x)^2 - 4(x + \delta x) + 1$

$\qquad y + \delta y = 4(x^2 + 2x\delta x + (\delta x)^2) - 4x - 4\delta x + 1$

$\qquad y + \delta y = 4x^2 + 8x\delta x + 4(\delta x)^2 - 4x - 4\delta x + 1$

But $y = 4x^2 - 4x + 1$

Subtracting these equations gives

$\qquad \delta y = 8x\delta x + 4(\delta x)^2 - 4\delta x$

$\qquad \dfrac{\delta y}{\delta x} = 8x + 4\delta x - 4$

Dividing both sides by δx and letting $\delta x \to 0$

$\qquad \dfrac{dy}{dx} = \underset{\delta x \to 0}{\text{limit}}\ \dfrac{\delta y}{\delta x} = 8x - 4$

(b) For an increasing function, $\dfrac{dy}{dx} > 0$

$\qquad$ Hence, $\quad 8x - 4 > 0$

$\qquad\qquad\qquad x > \dfrac{1}{2}$

4 Let $y = \dfrac{3}{x^4} + 4\sqrt{x}$

$\qquad y = 3x^{-4} + 4x^{\frac{1}{2}}$

$\qquad \dfrac{dy}{dx} = -12x^{-3} + 2x^{-\frac{1}{2}}$

$\qquad\quad = -\dfrac{12}{x^3} + \dfrac{2}{\sqrt{x}}$

Watch out

Not including the information about the limit will usually cost you a mark.

An increasing function means that the gradient is positive.

> Always check to see if a quadratic equation can divided by a number because if it can it will make factorisation easier.

Watch out

Make sure you give both coordinates and not just the x.

⑤ (a) Differentiating the equation of the curve to find the gradient gives:

$$\frac{dy}{dx} = 3x^2 - 6x - 9$$

At the stationary points, $\frac{dy}{dx} = 0$

Hence, $3x^2 - 6x - 9 = 0$.
Dividing both sides by 3 we obtain:
$$x^2 - 2x - 3 = 0.$$
Factorising gives $(x - 3)(x + 1) = 0$.
Hence $x = 3$ or -1
When $x = 3$, $y = (3)^3 - 3(3)^2 - 9(3) + 7 = -20$
When $x = -1$, $y = (-1)^3 - 3(-1)^2 - 9(-1) + 7 = 12$
Stationary points are at $(-1, 12)$ and $(3, -20)$
Differentiating again to find the nature of the stationary points:

$$\frac{d^2y}{dx^2} = 6x - 6$$

When $x = -1$, $\frac{d^2y}{dx^2} = 6(-1) - 6 = -12$

so $(-1, 12)$ is a maximum point.

When $x = 3$, $\frac{d^2y}{dx^2} = 6(3) - 6 = 12$

so $(3, -20)$ is a minimum point.

(b) To find the point where the curve cuts the y-axis we substitute $x = 0$ into the equation of the curve. Hence we have $y = (0)^3 - 3(0)^2 - 9(0) + 7 = 7$. So the curve cuts the y-axis at $(0, 7)$.

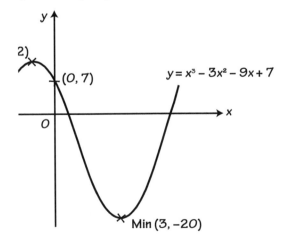

⑥ $y = 3x^{\frac{3}{2}} - \dfrac{32}{x}$

   $\ = 3x^{\frac{3}{2}} - 32x^{-1}$

$$\frac{dy}{dx} = \frac{9}{2}x^{\frac{1}{2}} + 32x^{-2}$$

$$= \frac{9}{2}\sqrt{x} + \frac{32}{x^2}$$

When $x = 4$, $\qquad \dfrac{dy}{dx} = \dfrac{9}{2}\sqrt{4} + \dfrac{32}{16}$

$$= 11$$

Using $m_1 m_2 = -1$

Gradient of normal $= -\dfrac{1}{11}$

When $x = 4$, $\qquad y = 3(4)^{\frac{3}{2}} - \dfrac{32}{(4)}$

$$= 24 - 8$$

$$= 16$$

Equation of normal with gradient $= -\dfrac{1}{11}$ and passing through $(4, 16)$ is

$$y - 16 = -\frac{1}{11}(x - 4)$$

$$11y - 176 = -x + 4$$

$$x + 11y - 180 = 0$$

7 (a) Length $= 8 - 2x$
Width $= 5 - 2x$
Height $= x$

$$V = (8 - 2x)(5 - 2x)x$$
$$= (40 - 26x + 4x^2)x$$
$$= 4x^3 - 26x^2 + 40x$$

(b) $\dfrac{dV}{dx} = 12x^2 - 52x + 40$

At max or min, $\quad \dfrac{dV}{dx} = 0$

$$12x^2 - 52x + 40 = 0$$
$$3x^2 - 13x + 10 = 0$$
$$(3x - 10)(x - 1) = 0$$

$x = \dfrac{10}{3}$ or $x = 1$

x cannot be $\dfrac{10}{3}$ as this would result in a negative width, so $x = 1$

When $x = 1$, $\qquad V = (8 - 2(1))(5 - 2(1))1 = 18\,\text{cm}^3$

$$\frac{d^2V}{dx^2} = 24x - 52$$

When $x = 1$, $\qquad \dfrac{d^2V}{dx^2} = 24(1) - 52 = -28$

As $\dfrac{d^2V}{dx^2} < 0$, maximum value of V is $18\,\text{cm}^3$

> If there are two answers, always ask yourself if both answers are acceptable. Here we can substitute both answers into the dimensions to check they are allowable.

> A negative value here means that $x = 1$ corresponds to a maximum value of V.

Exam practice answers

(8) (a) Let $y = 2x^5 + \dfrac{24}{x^2} - \sqrt[3]{x}$

$$y = 2x^5 + 24x^{-2} - x^{\frac{1}{3}}$$

$$\frac{dy}{dx} = 10x^4 - 48x^{-3} - \frac{1}{3}x^{-\frac{2}{3}}$$

(b) Let $y = \sqrt{x}\left(\sqrt[3]{x} - 2x + 1\right)$

$$= x^{\frac{5}{6}} - 2x^{\frac{3}{2}} + x^{\frac{1}{2}}$$

$$\frac{dy}{dx} = \frac{5}{6}x^{-\frac{1}{6}} - 3x^{\frac{1}{2}} + \frac{1}{2}x^{-\frac{1}{2}}$$

(9) $y = x^4 + x + 1$

$$\frac{dy}{dx} = 4x^3 + 1$$

> The tangent to the curve will have the same gradient as that of the curve.

When $x = 1$, $\qquad \dfrac{dy}{dx} = 4(1)^3 + 1 = 5$

Equation of tangent at $(1, 3)$ is
$$y - 3 = 5(x - 1)$$
$$y = 5x - 2$$

(10) (a) $y = x^3 - 3x^2 - 9x + 3$

$$\frac{dy}{dx} = 3x^2 - 6x - 9$$

At the stationary points, $\dfrac{dy}{dx} = 0$

$$3x^2 - 6x - 9 = 0$$
$$x^2 - 2x - 3 = 0$$
$$(x - 3)(x + 1) = 0$$

$x = -1$ or 3

Differentiating again to find the nature of the stationary points:

$$\frac{d^2y}{dx^2} = 6x - 6$$

When $x = -1$, $\qquad \dfrac{d^2y}{dx^2} = 6(-1) - 6 = -12$ so max. point.

When $x = 3$, $\qquad \dfrac{d^2y}{dx^2} = 6(3) - 6 = 12$ so min. point.

When $x = -1$, $\quad y = (-1)^3 - 3(-1)^2 - 9(-1) + 3 = 8$
When $x = 3$, $\qquad y = (3)^3 - 3(3)^2 - 9(3) + 3 = -24$
Max. point at $(-1, 8)$ and min. point at $(3, -24)$.

(b)

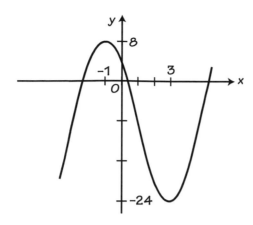

Topic 8

① $\int\left(4x^3 - \dfrac{4}{x^2} + 6\sqrt{x} - 1\right)dx = \int\left(4x^3 - 4x^{-2} + 6x^{\frac{1}{2}} - 1\right)dx$

$$= \dfrac{4x^4}{4} - \dfrac{4x^{-1}}{-1} + \dfrac{6x^{\frac{3}{2}}}{\frac{3}{2}} - x + c$$

$$= x^4 + 4x^{-1} + 4x^{\frac{3}{2}} - x + c$$

② $\dfrac{dy}{dx} = 3x^2 + 6x - 1$

$$y = \int(3x^2 + 6x - 1)\,dx$$

$$= \dfrac{3x^3}{3} + \dfrac{6x^2}{2} - x + c$$

$$= x^3 + 3x^2 - x + c$$

As the curve passes through (1, 7) these coordinates can be substituted into the equation for the curve to find c.

$$7 = 1^3 + 3(1)^2 - 1 + c$$
$$c = 4$$

Equation of curve C is $y = x^3 + 3x^2 - x + 4$

> We integrate the gradient function to go back to the equation of the curve.

③ (a) $\dfrac{dy}{dx} = x^2 + 2x - 8$

So, $y = \int(x^2 + 2x - 8)\,dx$

Integrating with respect to x gives:

$$y = \dfrac{x^3}{3} + \dfrac{2x^2}{2} - 8x + c$$

$$y = \dfrac{x^3}{3} + x^2 - 8x + c$$

> When you have the gradient, you can integrate this to take you back to the equation of the curve. You must remember to include the constant of integration, c.

By knowing the coordinates of a point on the curve, you can substitute these in to find the value of the constant of integration, .

If the question wanted just the x-coordinate, it would have specified it. If you are asked for the coordinates of the stationary points, you must give both the x and y-coordinates.

Now as the point $(3, 0)$ lies on the curve the coordinates will satisfy the equation of the curve. So,

$$0 = \frac{(3)^3}{3} + (3)^2 - 8(3) + c$$

Solving gives $c = 6$.

(b) At the stationary points, $\frac{dy}{dx} = 0$ so $x^2 + 2x - 8 = 0$.

Factorising the quadratic equation gives $(x - 2)(x + 4) = 0$.
Hence, $x = 2$ or -4

When $x = 2$, $\quad y = \frac{(2)^3}{3} + (2)^2 - 8(2) + 6 = -3\frac{1}{3}$

When $x = -4$, $\quad y = \frac{(-4)^3}{3} + (-4)^2 - 8(-4) + 6 = 32\frac{2}{3}$

(c) Finding the intercept on the y-axis by substituting $x = 0$ into the equation of the curve we have

$$y = \frac{(0)^3}{3} + (0)^2 - 8(0) + 6 = 6.$$

Adding the points to the sketch we have:

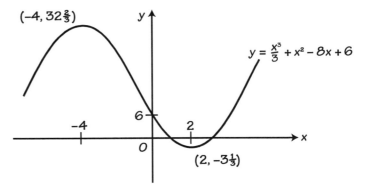

4 (a) Equating the y-values, we obtain $\quad 2x^2 = 12 - 2x$
$$2x^2 + 2x - 12 = 0$$
$$(2x + 6)(x - 2) = 0$$

Hence, $x = -3$ or $x = 2$
From the sketch, it can be seen that x is positive for point A, so $x = 2$.
When $x = 2$, $y = 12 - 2(2) = 8$.
A is the point $(2, 8)$
B is the point on the line where $y = 0$, so $0 = 12 - 2x$, giving $x = 6$.
B is the point $(6, 0)$.

(b) Area under the curve between $x = 0$ and $x = 2$ is given by:

$$= \int_0^2 2x^2 \, dx = \left[\frac{2x^3}{3}\right]_0^2$$

$$= \left[\left(\frac{2(2)^3}{3}\right) - \left(\frac{2(0)^3}{3}\right)\right]$$

$$= \frac{16}{3} - 0$$

$$= \frac{16}{3}$$

Area of triangle $= \frac{1}{2} \times 4 \times 8 = 16$

Total shaded area $= \frac{16}{3} + 16 = \frac{64}{3}$ square units

5 (a) $\int \left(2x^{\frac{3}{4}} + \frac{7}{x^{\frac{1}{2}}}\right) dx = \int \left(2x^{\frac{3}{4}} + 7x^{-\frac{1}{2}}\right) dx$

$$= \frac{2x^{\frac{7}{4}}}{\frac{7}{4}} + \frac{7x^{\frac{1}{2}}}{\frac{1}{2}} + c$$

$$= \frac{8}{7}x^{\frac{7}{4}} + 14x^{\frac{1}{2}} + c$$

(b) $6x - x^2 = 5$

$$x^2 - 6x + 5 = 0$$
$$(x - 5)(x - 1) = 0$$

$x = 1$ or 5

Hence A is (1, 5) and B is (5, 5)

Area under the curve between $x = 1$ and $x = 5$

$$= \int_1^5 (6x - x^2)\, dx$$

$$= \left[3x^2 - \frac{x^3}{3}\right]_1^5$$

$$= \left[\left(75 - \frac{125}{3}\right) - \left(3 - \frac{1}{3}\right)\right]$$

$$= \frac{92}{3}$$

> We equate the y-values to form an equation in x that can be solved to find the x-coordinates of the points of intersection.

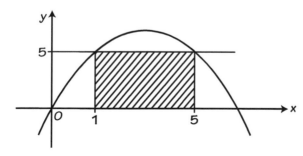

Area of shaded rectangle $= 4 \times 5 = 20$

Required area $= \frac{92}{3} - 20 = \frac{32}{3}$

6 $\int \sqrt[3]{x^2}\,(x-1)\,dx = \int x^{\frac{2}{3}}(x-1)\,dx$

$$= \int \left(x^{\frac{5}{3}} - x^{\frac{2}{3}}\right) dx$$

$$= \frac{x^{\frac{8}{3}}}{\frac{8}{3}} - \frac{x^{\frac{5}{3}}}{\frac{5}{3}} + c$$

$$= \frac{3}{8}x^{\frac{8}{3}} - \frac{3}{5}x^{\frac{5}{3}} + c$$

Topic 9

1 (a) $\overrightarrow{AB} = \overrightarrow{AO} + \overrightarrow{OB} = -\overrightarrow{OA} + \overrightarrow{OB} = -\mathbf{a} + \mathbf{b}$

Hence $\overrightarrow{AB} = \mathbf{b} - \mathbf{a} = 9\mathbf{i} - 2\mathbf{j} - (-2\mathbf{i} + 5\mathbf{j}) = 11\mathbf{i} - 7\mathbf{j}$

(b)

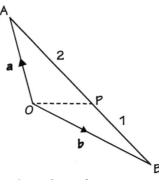

$\overrightarrow{OP} = \overrightarrow{OA} + \overrightarrow{AP}$

$$= \mathbf{a} + \frac{2}{2+1}\,\overrightarrow{AB}$$

$$= \mathbf{a} + \frac{2}{3}\left(11\mathbf{i} - 7\mathbf{j}\right)$$

$$= -2\mathbf{i} + 5\mathbf{j} + \frac{2}{3}\left(11\mathbf{i} - 7\mathbf{j}\right)$$

$$= -2\mathbf{i} + 5\mathbf{j} + \frac{22}{3}\mathbf{i} - \frac{14}{3}\mathbf{j}$$

$$= \frac{16}{3}\mathbf{i} - \frac{1}{3}\mathbf{j}$$

2 (a) (i) $\overrightarrow{AB} = \overrightarrow{AO} + \overrightarrow{OB} = -\overrightarrow{OA} + \overrightarrow{OB} = -\mathbf{i} - 3\mathbf{j} + 3\mathbf{i} + 7\mathbf{j} = 2\mathbf{i} + 4\mathbf{j}$
$\overrightarrow{BC} = \overrightarrow{BO} + \overrightarrow{OC} = -\overrightarrow{OB} + \overrightarrow{OC} = -3\mathbf{i} - 7\mathbf{j} + 4\mathbf{i} + 9\mathbf{j} = \mathbf{i} + 2\mathbf{j}$

(ii) Now $\overrightarrow{AB} = 2\overrightarrow{BC}$ so both have the same vector (i.e. $\mathbf{i} + 2\mathbf{j}$) part, so they are parallel and as they both pass through B, points A, B and C lie on a straight line.

This means AB is twice as long as BC.

(b) $\overrightarrow{AB} = 2\overrightarrow{BC}$
AB:BC = 2:1

3 The point P dividing AB in the ratio $\lambda:\mu$ has position vector

$$\overrightarrow{OP} = \frac{\mu\mathbf{a} + \lambda\mathbf{b}}{\mu + \lambda}$$

As $\lambda = 8$ and $\mu = 3$,

$$\overrightarrow{OP} = \frac{\mu\mathbf{a} + \lambda\mathbf{b}}{\mu + \lambda}$$

$$= \frac{3(-12\mathbf{i} - 5\mathbf{j}) + 8(10\mathbf{i} + 6\mathbf{j})}{8 + 3}$$

$$= \frac{-36\mathbf{i} - 15\mathbf{j} + 80\mathbf{i} + 48\mathbf{j}}{11}$$

$$= \frac{44\mathbf{i} + 33\mathbf{j}}{11}$$

$$= 4\mathbf{i} + 3\mathbf{j}$$

4 (a) (i) $\overrightarrow{OB} = \overrightarrow{OA} + \overrightarrow{AB}$

$\qquad\qquad = \mathbf{a} + \mathbf{c}$

(ii) $\overrightarrow{DE} = \overrightarrow{DB} + \overrightarrow{BE}$

Now $\qquad \overrightarrow{EB} = \frac{1}{3}\overrightarrow{AB} = \frac{1}{3}\mathbf{c}$

$$\overrightarrow{BE} = -\frac{1}{3}\mathbf{c} \text{ and } \overrightarrow{DB} = \frac{1}{2}\mathbf{a}$$

$$\overrightarrow{DE} = \overrightarrow{DB} + \overrightarrow{BE} = \frac{1}{2}\mathbf{a} - \frac{1}{3}\mathbf{c}$$

(b) $\overrightarrow{CE} = \overrightarrow{CB} + \overrightarrow{BE} = \mathbf{a} - \frac{1}{3}\mathbf{c}$

As the vectors for DE and CE are not equal (or a multiple of each other) the lines DE and CE are not parallel.

5 (a)

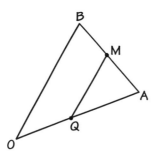

$$\overrightarrow{AB} = \overrightarrow{AO} + \overrightarrow{OB} = -\overrightarrow{OA} + \overrightarrow{OB} = -10\mathbf{i} - 2\mathbf{j} + 8\mathbf{i} + 8\mathbf{j}$$
$$= -2\mathbf{i} + 6\mathbf{j}$$

$$\overrightarrow{AM} = \frac{1}{2}\overrightarrow{AB} = \frac{1}{2}(-2\mathbf{i} + 6\mathbf{j}) = -\mathbf{i} + 3\mathbf{j}$$

$$\overrightarrow{QM} = \overrightarrow{QA} + \overrightarrow{AM} \text{ where } \overrightarrow{QA} = \frac{1}{2}\overrightarrow{OA}$$
$$= 5\mathbf{i} + \mathbf{j} - \mathbf{i} + 3\mathbf{j}$$
$$= 4\mathbf{i} + 4\mathbf{j}$$

(b) $\vec{OB} = 8\mathbf{i} + 8\mathbf{j}$ and $\vec{QM} = 4\mathbf{i} + 4\mathbf{j}$

Hence $\vec{OB} = 2(4\mathbf{i} + 4\mathbf{j}) = 2\vec{QM}$

As both lines have the same vector (i.e. $4\mathbf{i} + 4\mathbf{j}$) it means that QM and OB are parallel.

Topic 10

1 Each student is given a unique number from 1 to 1500 inclusive. A random number generator such as a computer program or on a calculator is used to create random numbers from 1 to 1500. If a number has already been picked, the repeated number is ignored, and a new random number is picked. Once 50 random numbers have been picked then numbers are matched to the names and then each student is selected.

2 (a) Mean $= \dfrac{2 + 4 + 0 + 1 + 3}{5} = 2$

(b) (i) Sampling interval $= \dfrac{\text{population}}{\text{sample size}} = \dfrac{25}{5} = 5$

(ii) 4, 5, 1, 2, 1

(iii) Mean $= \dfrac{4 + 5 + 1 + 2 + 1}{5} = 2.6$

(c) The systematic sample as it is more random and it uses numbers throughout the distribution and not simply the first 5 values which may not be typical of the rest.

> The random number 2 means that you count along to the second value in the list. This is then the first data item in the sample. Now count along five numbers and this gives the second data value. This is repeated until the 5 data values are obtained.

3 (a) (i) Taking a sample from the population that is easy to take. For example, the first so many numbers in a list, the first so many customers visiting a shop, etc.

(ii) Mean $= \dfrac{\substack{32\,000 + 28\,000 + 20\,000 + 63\,000 + 18\,000 + 29\,000 \\ + 50\,000 + 18\,000 + 74\,000 + 48\,000}}{10}$

$= £38\,000$

(b) (i) sampling interval $= \dfrac{\text{population}}{\text{sample size}} = \dfrac{30}{6} = 5$

(ii) 63\,000, 74\,000, 94\,000, 85\,000, 78\,000, 89\,000

(iii) Mean $= \dfrac{63\,000 + 74\,000 + 94\,000 + 85\,000 + 78\,000 + 89\,000}{6}$

$= £80\,500$

(iv) The systematic sample just happens to pick some very high numbers as part of the sample. These numbers are members of the sample but just happen to occur every fifth number starting from the 4th number.

4 (a) The population is all the members of the set being studied so here it will be all the students in Jack's school.
The sample is a smaller subset of the population used to draw conclusions about the population.

(b) The sample is too small as it is only 30 out of the population of 1200.
He has only asked students in his class so they will all be a certain age and may not be representative of students of different ages.

(c) (i) There is a wide difference in the numbers so this is likely to be a correct inference.

(ii) This is an inference that should not be made as, although the numbers are the same for this sample, if other samples were used they would probably be different.

Topic 11

1 (a) Strong negative correlation.
Resting heart rate decreases as the number of hours of exercise increases.

(b) (i) The gradient represents the decrease in resting heart rate per hour of exercise per week.

(ii) This is the resting heart rate for a person who does no hours of exercise per week.

(iii) This is the last reading there is data for. The line could behave in a different way past this last point.

(iv) The amount of regular exercise is shown to lower your heart rate so it is causal.

(v) No it could not be used as 5 hours per day is 35 hours per week and the diagram only goes as high as 8.8 hours per week. There is no evidence that the line behaves in the same way past this point.

2 (a) (i) Strong positive correlation.

(ii) As the chemistry marks increased, so did the physics marks.

(b) (i) $IQR = 68.8 - 26.3 = 42.5$
Outlier would be larger than $Q_3 + 1.5 \times IQR$
$= 68.8 + 42.5 =$ over 100
As 90 is not over 100 it is not an outlier.

(ii) The median mark for physics at 39 is higher than 37.5 for chemistry.
The range for chemistry is 84 whilst that for physics is 80 and the IQR for chemistry is 56 whilst for physics it is 42.5. This means that the physics marks show smaller variation and are more consistent than the chemistry marks. So overall, they did better at physics and the physics marks showed less spread.

Exam practice answers

3 (a) The median is the $\frac{n+1}{2}$ th value $= \frac{30+1}{2} = 15.5$ so it is the mean of the 15th and 16th values in the list.

$$\text{Median age} = \frac{3+3}{2} = 3 \text{ years}$$

(b) Lower quartile $= \frac{n+1}{4}$ th value $= \frac{30+1}{4} = 7.75$ the mean of the 7th and 8th values.

$$\text{Lower quartile} = \frac{2+2}{2} = 2$$

Upper quartile $= \frac{3(n+1)}{4} = \frac{3 \times 31}{4} = 23.25$ the mean of the 23rd and 24th values.

$$\text{Upper quartile} = \frac{6+6}{2} = 6$$

(c) Larger outliers are values larger than $Q_3 + 1.5 \times IQR$
IQR = upper quartile − lower quartile = 6 − 2 = 4
$$Q_3 + 1.5 \times IQR = 6 + 1.5 \times 4 = 12$$
The value of 20 is an outlier as 20 > 12
Small outliers are values smaller than $Q_3 - 1.5 \times IQR$
$$Q_3 - 1.5 \times IQR = 6 - 1.5 \times 4 = 0$$
There are no values smaller than 0.
The only outlier is the value 20 years.

4 (a) Positive correlation.
The greater the traffic volume per day the greater the carbon monoxide level in parts per million.

(b) The gradient represents the rate of increase in carbon monoxide per vehicle.

(c) This is the carbon monoxide level when there are no cars per day.

(d) The scatter diagram has a maximum number of vehicles per day of 470. The traffic volume on a motorway is much greater than this so extrapolation would be needed. There is no reason to believe that the graph would behave in the same way as the speeds and type of vehicles would not be the same.

5 (a) Chemistry IQR = 79 − 22 = 57
Physics IQR = 70.5 − 24.5 = 46

(b) Chemistry range = 90 − 5 = 85
Physics range = 90 − 10 = 80
Chemistry median = 37.5
Physics median = 39
The spread of the marks is greater for chemistry compared to physics as shown by the larger IQR and range. The median mark for physics (39) is slightly greater than that for chemistry (37.5). To conclude the marks for physics were higher and more consistent.

6 (a) 40 miles per gallon.

(b) Negative correlation.

(c) Causal, as it is known that cars with larger engines do fewer miles per gallon.

(d) Interpolation – using the line of best fit within the range of the first and last points.
Extrapolation – using the line of best fit outside the range of the first and last points.

(e) 5000 cm^3 is beyond the last point on the graph. There is no evidence that the line will behave as it did previously beyond this point.

7 (a) (i) Lower quartile = $\frac{n+1}{4}$ th value = $\frac{30+1}{4}$ = the mean of the 7th and 8th values.

$$Q_1 = \frac{56+63}{2} = 59.5$$

Upper quartile = $\frac{3(n+1)}{4} = \frac{3 \times 31}{4}$ = 23.25 the mean of the 23rd and 24th values.

$$Q_3 = \frac{257+260}{2} = 258.5$$

(ii) IQR = 258.5 − 59.5 = 199
Larger outliers are values larger than
$Q_3 + 1.5 \times$ IQR = 258.5 + 1.5 × 199 = 557
Hence 600 is an outlier.
Smaller outliers are values smaller than
$Q_1 − 1.5 \times$ IQR = 59.5 − 1.5 × 199 = −239
Hence there are no small outliers.

(b) (i) The removal of three large data values will reduce the upper quartile and make the inter-quartile range lower. It will also reduce the lower quartile owing to there being fewer numbers in the data set.

(ii) The removal of three large data values will move the median to the left of the original median so it will go lower.

(iii) The mean uses all the data values to obtain the total. Removal will lower this total so the mean will be lower.

8

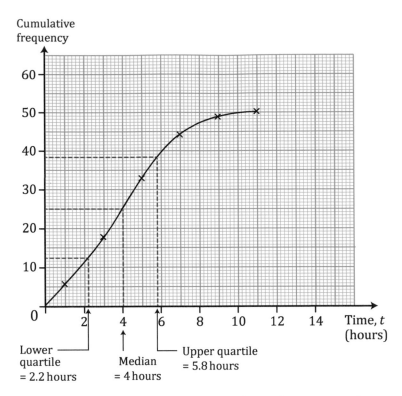

From the cumulative frequency graph for the boys, we have:

Lower quartile = 2.2 hours

Median = 4.0 hours

Upper quartile = 5.8 hours

Range = 8.5 hours

Interquartile range = 5.8 − 2.2 = 3.6 hours

From the box and whisker diagram for the girls, we have:

Lower quartile = 1.6 hours

Median = 4.6 hours

Upper quartile = 6 hours

Interquartile range = 6 − 1.6 = 4.4 hours

Range = 8 − 0.2 = 7.8 hours

The average time was lower for the boys than for the girls as shown by the medians of 4 hours and 4.6 hours.

The overall spread of times was greater for the boys as shown by the greater range.

9 $\dfrac{\sum x_i}{n} = \dfrac{92}{20} = 4.6$

Variance $= \dfrac{\sum x_i^2}{n} - \left(\dfrac{\sum x_i}{n}\right)^2 = \dfrac{476}{20} - 4.6^2 = 2.64 = 2.6 \text{ (2 s.f.)}$

Standard deviation $= \sqrt{\text{variance}} = \sqrt{2.64} = 1.6 \text{ (2 s.f.)}$

10 (a) Positive correlation – the greater the amount of carbohydrate the greater the number of calories.
   (b) It is because both axes do not start from zero. The steepness of the line has been overemphasised.
   (c) It is causal as it is well known that carbohydrates contain lots of calories.

Topic 12

1 (a) $P(A \cup B) = P(A) + P(B) - P(A \cap B)$
        As the events are independent, $P(A \cap B) = P(A) \times P(B)$
        Hence $P(A \cup B) = P(A) + P(B) - P(A) \times P(B)$
        $\qquad\qquad\qquad = 0.7 + 0.4 - 0.7 \times 0.4$
        $\qquad\qquad\qquad = 0.82$

   (b) (i)

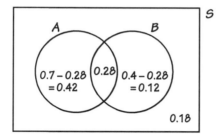

P(exactly one of A and B will occur) $= 0.42 + 0.12 = 0.54$
   (ii) P(neither A nor B will occur) $= 0.18$

2 (a)

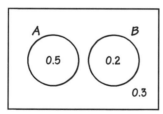

$\qquad P(A \cup B) = P(A) + P(B)$
$\qquad\qquad 0.7 = 0.5 + P(B)$
$\qquad\qquad P(B) = 0.2$

   (b) $P(A \cap B) = P(A) \times P(B)$
       $P(A \cup B) = P(A) + P(B) - P(A \cap B)$
       Hence, $P(A \cup B) = P(A) + P(B) - P(A) \times P(B)$
       $\qquad\qquad 0.7 = 0.5 + P(B) - 0.5P(B)$
       $\qquad\qquad 0.2 = 0.5P(B)$
       $\qquad\qquad P(B) = 0.4$

Drawing a Venn diagram and writing in the relevant probabilities makes this question easier to answer.

This represents the probability of A but not B or the probability of not A but B.

As events A and B are mutually exclusive, A can occur or B can occur but not both. Hence there is no overlap between sets A and B.

③ (a) $P(A \cup B) = P(A) + P(B) - P(A) \times P(B)$

$0.4 = 0.2 + P(B) - 0.2P(B)$

$0.2 = 0.8P(B)$

$P(B) = 0.25$

(b) $P(A \cap B) = P(A) \times P(B) = 0.2 \times 0.25 = 0.05$

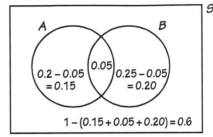

$P(\text{exactly one of the two events occurs}) = 0.15 + 0.20$

$= 0.35$

(c) $P(\text{given that exactly one of the two events occurs, event } A \text{ occurs})$

$= \dfrac{\text{Probability only } A \text{ occurs}}{\text{Probability of one event ocurring}} = \dfrac{0.15}{(0.15 + 0.20)} = \dfrac{3}{7}$

④ (a) As the events are independent, $P(A \cap B) = P(A) \times P(B)$

$= 0.6 \times 0.3 = 0.18$

We can now draw the Venn diagram:

> Note that you could have used the formula
> $P(A \cup B) = P(A) + P(B) - P(A \cap B)$
> instead of drawing the Venn diagram.

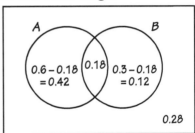

$P(A \cup B) = 0.42 + 0.18 + 0.12 = 0.72$

(b) $P(A \cup B') = 0.42 + 0.18 + 0.28 = 0.88$

⑤ (a) As the events are independent, $P(A \cap B) = P(A) \times P(B)$

$= 0.7 \times 0.4 = 0.28$

$P(A \cup B) = P(A) + P(B) - P(A \cap B) = 0.7 + 0.4 - 0.28 = 0.82$

(b) (i) You can draw a Venn diagram to show the required region (shown shaded). This represents the probability of A but not B or the probability of not A but B.

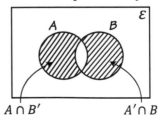

$A \cap B'$ $A' \cap B$

Now $\quad$ P$(B') = 1 - P(B) = 1 - 0.4 = 0.6$
$\qquad$ P$(A') = 1 - P(A) = 1 - 0.7 = 0.3$
$\qquad$ Required probability $= P(A \cap B') + P(A' \cap B)$
$\qquad\qquad\qquad\qquad\qquad = 0.7 \times 0.6 + 0.3 \times 0.4$
$\qquad\qquad\qquad\qquad\qquad = 0.42 + 0.12$
$\qquad\qquad\qquad\qquad\qquad = 0.54$
$\quad$ (ii) $\;$ P$(A \cup B)' = 1 - P(A \cup B) = 1 - 0.82 = 0.18$

6 $\quad$ (a) $\;$ P$(A \cup B) = P(A) + P(B)$
$\qquad\qquad\qquad\quad = 0.4 + 0.2$
$\qquad\qquad\qquad\quad = 0.6$

$\quad$ (b) $\;$ P$(A \cup B) = P(A) + P(B) - P(A \cap B)$
$\qquad\qquad\qquad\quad = 0.4 + 0.2 - P(A) \times P(B)$
$\qquad\qquad\qquad\quad = 0.6 - 0.4 \times 0.2 = 0.52$

$\quad$ (c) $\;$ P$(A \cup B) = 0.4$

7 $\quad$ (a) $\;$ Let $x =$ number of students who take both French and Spanish.

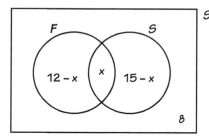

$\qquad\qquad 12 - x + x + 15 - x + 8 = 30$
$\qquad\qquad\qquad\qquad\qquad 35 - x = 30$
$\qquad\qquad\qquad\qquad\qquad\qquad x = 5$

$\qquad$ Probability of taking French and Spanish $= \dfrac{5}{30} = \dfrac{1}{6}$

$\quad$ (b) $\;$ Number taking French but not Spanish $= 12 - x = 12 - 5 = 7$

$\qquad$ Probability of taking French but not Spanish $= \dfrac{7}{30}$

8 $\quad$ (a) $\;$ The set of students who do not take Spanish, French or German.

$\quad$ (b) $\;$ (i) $\;$ P(French only) $= \dfrac{10}{42} = \dfrac{5}{21}$

$\qquad\quad$ (ii) $\;$ P(French or German or both) $= \dfrac{22}{42} = \dfrac{11}{21}$

$\quad$ (c) $\;$ If the events are independent,
$\qquad\qquad$ P(French) $\times$ P(Spanish) $=$ P(French and Spanish)

$\qquad\qquad$ P(French) $\times$ P(Spanish) $= \dfrac{18}{42} \times \dfrac{18}{42} = \dfrac{9}{49} = 0.18 \ldots$

$\qquad\qquad$ P(French and Spanish) $= P(F \cap S) = \dfrac{6}{42} = \dfrac{1}{7} = 0.14 \ldots$

$\quad$ Hence the events are not independent.

Topic 13

A binomial distribution is used here as you know the total number of parts, n and the probability of a part being faulty, p.

1 (a) Let the random variable X be the number of parts that fail in the first year.
$$X \sim B(n, p)$$
$$X \sim B(20, 0.05)$$

Using $P(X = x) = \binom{n}{x} p^x (1 - p)^{n-x}$,

This formula is looked up in the formula booklet.

with $x = 1$, $n = 20$ and $p = 0.05$ we obtain:
$$P(X = 1) = \binom{20}{1} 0.05^1 (1 - 0.05)^{20-1}$$
$$= \binom{20}{1} 0.05^1 (0.95)^{19}$$
$$= 0.3773536$$
$$= 0.3774 \text{ (correct to 4 s.f.)}$$

Remember to always define the what you are using as the random variable X. Also write the shorthand notation that tells the examiner which distribution you are using and the parameters of the distribution

Watch out

You need to be careful that you read the question carefully. Some students found $P(X \geq 4)$ by mistake.

(b) $P(X > 4) = 1 - P(X \leq 4)$
Use the Binomial CD function on the calculator with $x = 4$, $N = 20$ and $p = 0.05$ to find $P(X < 4)$.
$$= 1 - 0.9974$$
$$= 0.0026$$

The Poisson distribution is used to model the situation as you are modelling an event that occurs randomly within a certain interval of time. Notice that the mean rate of sales (i.e. 6 per week) is given.

2 (a) (i) Let the random variable X be the number of cars sold in a week.
$$X \sim Po(\lambda)$$
$$X \sim Po(6)$$
$$P(X \geq 4) = 1 - P(X \leq 3)$$
$$= 1 - 0.1512$$
$$= 0.8488$$

The tables or Poisson CD function on a calculator give the probability less than or equal to a certain value X. Hence we need to perform this calculation.

We use the Poisson CD function on the calculator with $x = 3$ and $\lambda = 6$ to work out $P(X \leq 3)$.

Alternatively you could use the Poisson CD function on the calculator by using:
$P(X = 6) = P(X \leq 6) -$
$P(X \leq 5)$

(ii) We need to find $P(X = 6)$.
The Poisson formula is looked up: $P(X = x) = e^{-\lambda} \dfrac{\lambda^x}{x!}$
Now $x = 6$ and $\lambda = 6$.

Hence, $P(X = 6) = e^{-6} \dfrac{6^6}{6!}$
$$= 0.1606$$

(b) (i) Let the random variable X be the number of motorbikes sold in a week.

$$X \sim Po(\lambda)$$
$$X \sim Po(1.12)$$

We need to find $P(X = 2)$.

The Poisson formula is looked up: $P(X = x) = e^{-\lambda}\dfrac{\lambda^x}{x!}$

Now $x = 2$ and $\lambda = 1.12$.

Hence, $P(X = 2) = e^{-1.12}\dfrac{1.12^2}{2!}$

$= 0.2046$

> Alternatively, you can use the Poisson CD function and:
>
> $P(X = 2) = P(X \le 2) - P(X \le 1)$

(ii) $P(X \ge 2) = 1 - P(X \le 1)$

$= 1 - 0.6917$

$= 0.3083$

3 (a) If you know the mean probability of an event happening per unit time, area, etc., and you are asked to find the probability of a certain number of events happening (i.e. n) in a given time or area, then the Poisson distribution is used.

(b) Let the random variable X be the number of newts in a random $2\ m^2$ area of the pond.

Mean number of newts per $2\ m^2 = 0.5 \times 2 = 1$

$$X \sim Po(\lambda)$$
$$X \sim Po(1)$$
$$P(X \le 3) = 0.9810$$

> Note that 'at most 3' means less or equal to 3.

> The Poisson CD function is used on the calculator with $x = 3$ and $\lambda = 1$

(c) We need to find $P(X = 2)$.

The Poisson formula is looked up: $P(X = x) = e^{-\lambda}\dfrac{\lambda^x}{x!}$

Now $x = 2$ and $\lambda = 1$

Hence, $P(X = 2) = e^{-1}\dfrac{1^2}{2!}$

$= 0.1839$

> Alternatively, you can use the Poisson CD function and:
>
> $P(X = 2) = P(X \le 2) - P(X \le 1)$

Let the random variable Y be the number of chosen areas where there are exactly two newts.

$$Y \sim B(n, p)$$
$$Y \sim B(4, 0.1839)$$

> Here we want to find the probability of two occurrences from a total of four. We now need to use the Binomial probability distribution.

Using $P(Y = x) = \binom{n}{x}p^x(1 - p)^{n-x}$,

with $x = 2$, $n = 4$ and $p = 0.1839$ we obtain:

$$P(Y = 2) = \binom{4}{2}0.1839^2(1 - 0.1839)^2$$

$$= 0.1351$$

> **Watch out**
>
> We need to use the probability calculated using the Poisson distribution for p for the model using the binomial distribution.

Exam practice answers

④ (a) The exact probability of an event is unknown (as only the mean rate can be worked out). Also, we are not asked to find the probability of an event happening a certain number of times x out of a total number of times n.

(b) (i) Let the random variable X be the number of mistakes per page.

$$X \sim \text{Po}(\lambda)$$

Average rate of mistakes, $\lambda = 2$ per page

$$X \sim \text{Po}(2)$$

We need to find $P(X = 0)$.

The Poisson formula is looked up: $P(X = x) = e^{-\lambda}\dfrac{\lambda^x}{x!}$

Now $x = 0$ and $\lambda = 2$.

Hence, $P(X = 0) = e^{-2}\dfrac{2^0}{0!} = 0.1353$

(ii) $\lambda = 2$ and $x = 4$

$$P(X = 4) = e^{-2}\dfrac{2^4}{4!} = 0.0902$$

> Notice that this is a rate so we use the Poisson distribution.

> Note here that any number to the power zero is one and also that $0! = 1$

⑤ (a) (i) Let the random variable X be the number of errors on each page.

$$X \sim \text{Po}(\lambda)$$
$$X \sim \text{Po}(0.95)$$

The Poisson formula is looked up: $P(X = x) = e^{-\lambda}\dfrac{\lambda^x}{x!}$

$\lambda = 0.95$ and $x = 0$

$$P(X = 0) = e^{-0.95}\dfrac{(0.95)^0}{0!} = 0.3867$$

(ii) $\lambda = 0.95$ and $x = 3$

$$P(X = 3) = e^{-0.95}\dfrac{(0.95)^3}{3!} = 0.0553$$

$$P(X = 4) = e^{-0.95}\dfrac{(0.95)^4}{4!} = 0.0131$$

Probability of 3 or 4 errors = 0.0553 + 0.0131 = 0.0684

(b) (i) P(no errors on a page) = 0.3867
P(no errors on 4 pages) = 0.3867^4
= 0.0224

(ii) P(the first error occurs on the third page)
= P(no errors on 1st page) × P(no errors on 2nd page)
× P(an error on 3rd page)
= 0.3867 × 0.3867 × (1 − 0.3867)
= 0.0917

> Remember that $0! = 1$.

> Note that the 'AND' rule for probability is used here.

⑥ (a) (i) Let the random variable X be the number of accidents per week.

$$X \sim \text{Po}(\lambda)$$
$$X \sim \text{Po}(2.75)$$

The Poisson formula is looked up: $P(X = x) = e^{-\lambda}\dfrac{\lambda^x}{x!}$

$\lambda = 2.75$ and $x = 4$

$$P(X = 4) = e^{-2.75}\frac{(2.75)^4}{4!} = 0.1523$$

Notice that a mean rate is given (i.e. 2.75 accidents per week) so we need to use the Poisson distribution.

(ii) $\lambda = 2.75$ and $x > 2$

$P(X > 2) = 1 - P(X \le 2)$

Here we use the Poisson CD function on the calculator to find $P(X \le 2)$ with $x = 2$ and $\lambda = 2.75$.

$P(X \le 2) = 0.4815$

Hence $P(X > 2) = 1 - 0.4815 = 0.5185$

Watch out

More than 2 does not include 2.

(b) The Poisson formula is looked up: $P(X = x) = e^{-\lambda}\dfrac{\lambda^x}{x!}$

$\lambda = 2.75$ and $x = 0$

$$P(X = 0) = e^{-2.75}\frac{(2.75)^0}{0!} = 0.0639$$

We first find the probability that there were no accidents in a randomly picked week.

Here we use the AND rule to find the probability.

Probability of no accidents over 3 weeks $= 0.0639^3$

$= 0.0003$ (1 s.f.)

7 (a) Let the random variable X be the number of holes in one putt.

$X \sim B(n, p)$

$X \sim B(10, 0.7)$

$P(X \ge 5) = 1 - P(X \le 4)$

$= 1 - 0.0473$

$= 0.9527$

We use binomial as the probability is known and we want to find the probability of x occurrences out of a total of n.

(b) $P = 0.9527^4$

$= 0.8238$

We have used a calculator to work out $P(X \le 4)$. Use the binomial CD function entering $x = 4$, $N = 10$ and $p = 0.7$.

(c) The probability of holing a ball stays constant. Or the probability of holing a ball with one putt is independent of holing balls with other putts.

The AND rule for probabilities is used here. Each occasion is assumed to have the same probability.

8 (a) The exact probability of an event is unknown as only the mean rate is known. Poisson is used to model events occurring over a period of time.

(b) (i) The Poisson formula is looked up: $P(X = x) = e^{-\lambda}\dfrac{\lambda^x}{x!}$

$\lambda = 24$ and $x = 20$

$$P(X = 20) = e^{-24}\frac{(24)^{20}}{20!} = 0.0624$$

The Poisson CD on the calculator is used to arrive at this value using $x = 24$ and $\lambda = 24$.

(ii) $P(X \le 24) = 0.5540$

(c) $P(30 \le X \le 40) = P(X \le 40) - P(X \le 29)$

$= 0.9990 - 0.8679$

$= 0.1311$

Exam practice answers

⑨ (a) Poisson is used because the constant mean arrival rate can be found, and the arrivals occur randomly, and we are modelling the number of events over a certain time period.

(b) 20 customers per hour are divided between four staff.

Mean for Jane $= \dfrac{20}{4} = 5$ per hour.

Mean rate in 30 min = 2.5 so $\lambda = 2.5$.

Let the random variable X be the number of customers Jane deals with in a 30-minute period.

$$X \sim \text{Po}(\lambda)$$
$$X \sim \text{Po}(2.5)$$

The Poisson formula is looked up: $P(X = x) = e^{-\lambda}\dfrac{\lambda^x}{x!}$

$\lambda = 2.5$ and $x = 4$

$$P(X = 4) = e^{-2.5}\frac{(2.5)^4}{4!} = 0.1336$$

We now use the Poisson CD function on the calculator. We are looking for a probability of 0.9 (or as close as we can get to it) with $x = 5$ to find a value for the mean λ. Note that we are going to have to use trial and improvement to find this value.

(c) We know $P(X > 5) = 0.1$

Hence $P(X \leq 5) = 1 - P(X > 5)$
$= 1 - 0.1$
$= 0.9$

Using $x = 5$, we vary λ until we obtain a probability of as near to 0.9 correct to one decimal place as possible.

When $x = 5$ and $\lambda = 3$, $\quad P(X \leq 5) = 0.9161$ which is too high.
When $x = 5$ and $\lambda = 4$, $\quad P(X \leq 5) = 0.7851$ which is too low.
When $x = 5$ and $\lambda = 3.2$, $P(X \leq 5) = 0.8946$ which is too low.
When $x = 5$ and $\lambda = 3.1$, $P(X \leq 5) = 0.9057$ which is too high.
$\lambda = 3.2$, gives the value of $P(X \leq 5)$ nearest to 0.9.

Hence the mean $\lambda = 3.2$.

Time interval $= \dfrac{3.2}{5} \times 60 = 38.4$ minutes

⑩ (a) The Binomial distribution.

(b) There are only two possible outcomes – hitting the bullseye or not hitting the bullseye.
Hitting the bullseye with one of the darts is independent of hitting the bullseye with any other dart.

Note that this could also be written as 'the probability of hitting the bullseye is fixed at 0.2'.

(c) Let the random variable X be the number of bullseyes obtained.

$$X \sim \text{B}(n, p)$$
$$X \sim \text{B}(10, 0.2)$$
$$P(X \geq 3) = 1 - P(X \leq 2)$$
$$= 1 - 0.6778$$
$$= 0.3222$$

Note this is the AND law for probability as you want a bull and a bull and a bull.

(d) P (hitting at least 3 times each on 3 separate occasions)
$$= 0.3222^3$$
$$= 0.0334$$

Topic 14

① (a) Under the null hypothesis, $\mathbf{H}_0$, $X \sim \text{B}(30, 0.2)$.
Significance level, α, $= \text{P}(X \le 3)$
$= 0.1227$

As the critical region is $X \le 3$, the significance level is $\text{P}(X \le 3)$. We use the binomial CD function on the calculator with $x = 3$, $N = 30$ and $p = 0.2$.

(b) A type II error is the error that occurs when you retain (i.e. fail to reject) a null hypothesis that is actually false.

(c) X is $\text{B}(30, 0.15)$
Type II error probability $= \text{P}(X \ge 4 \text{ given that } X \text{ is B}(30, 0.15))$
$= 1 - \text{P}(X \le 3 \text{ given that } X \text{ is B}(30, 0.15))$
$= 1 - 0.3217$
$= 0.6783$

As the critical region is $X \le 3$, $\text{P}(X \le 3)$ is the probability of rejecting the null hypothesis, so the probability of not rejecting it is $\text{P}(X \ge 4)$. Hence under the null hypothesis, the probability of failing to reject the null hypothesis when it is false is $\text{P}(X \ge 4)$.

② (a) (i) $\text{P}(X = x) = \binom{n}{x} p^x (1 - p)^{n-x}$

$\text{P}(X = 5) = \binom{55}{5}(0.3)^5 (1 - 0.3)^{55-5}$
$= 0.00015$

(ii) $\text{P}(10 \le X \le 11) = \text{P}(X = 10) + \text{P}(X = 11)$
$= \binom{55}{10}(0.3)^{10} (1 - 0.3)^{55-10} + \binom{55}{11}(0.3)^{11} (1 - 0.3)^{55-11}$
$= 0.0185 + 0.0324$
$= 0.0509$

(b) If p is the probability of a student drinking 14 or more units per week.
Null hypothesis is $\qquad \mathbf{H}_0 : p = 0.3$
Alternative hypothesis is $\quad \mathbf{H}_1 : p > 0.3$
The random variable, X, is the number of students who drink 14 or more units of alcohol per week.
$$X \sim \text{B}(20, 0.3)$$
$$\text{P}(X \ge 8) = 1 - \text{P}(X < 8) = 1 - \text{P}(X \le 7)$$
We now use the binomial CD on the calculator (or you could use tables) with $x = 7$, $p = 0.3$ and $N = 20$.
$$\text{P}(X \ge 8) = 1 - 0.7723$$
$$= 0.2277$$
Now $\qquad 0.2277 > 0.05$

This is the p-value which is now compared with the level of significance of 0.05.

Hence, we fail to reject the null hypothesis. This means there is not enough evidence to accept James' proposition that the percentage drinking 14 or more units of alcohol is higher than the original research.

③ (a) (i) X is $\text{B}(20, 0.5)$
$$\text{P}(X \ge 14) = 1 - \text{P}(X \le 13)$$
$$= 1 - 0.9423$$
$$= 0.0577$$
Significance level $= 0.0577$

Note that $X \ge 14$ is the critical region. Hence $\text{P}(X \ge 14)$ is the significance level.

Exam practice answers

If you want a single probability you can use the binomial formula which is obtained from the formula booklet.

The p-value is compared with the significance level previously calculated and as it is higher we fail to reject the null hypothesis.

(ii) $X \sim B(20, 0.7)$
$$P(X \geq 14) = 1 - P(X \leq 13)$$
$$= 1 - 0.3920$$
$$= 0.6080$$

(b) (i) $P(X = 28) = \binom{50}{28}(0.5)^{28}(1 - 0.5)^{50-28}$
$$= \binom{50}{28}(0.5)^{28}(0.5)^{22}$$
$$= 0.0788$$

(ii) $0.0788 > 0.0577$
There is not enough evidence to support Dafydd's theory.

4 (a) (i) If p is the probability of Bryn obtaining a head:
Null hypothesis is $\quad$ $H_0 : p = 0.5$
Alternative hypothesis is $\quad$ $H_1 : p > 0.5$

(ii) The test statistic, X, is the number of times a head is thrown.

(b) The critical region is the set of all the values that would cause the null hypothesis H_0 to be rejected.

(c) $P(X \geq 8) = 1 - P(X \leq 7) = 1 - 0.9453 = 0.0547$
Now $0.0547 > 0.05$ so this would not cause the null hypothesis to be rejected and $X \geq 8$ is not the critical region.
$$P(X \geq 9) = 1 - P(X \leq 8) = 1 - 0.9893 = 0.0107$$
$0.0107 < 0.05$ so this would cause the null hypothesis to be rejected.
Hence, the critical region is $X \geq 9$

(d) The 8 heads obtained is outside the critical region so there is insufficient evidence to reject the null hypothesis and there is insufficient evidence to conclude that the coin is biased.

5 Using the binomial CD function.
$N = 20$ and $p = 0.3$. However, we don't know the value of x as we are trying to find which is the first value of x that will give a probability that exceeds 0.05.
If you find $N \times p$ it gives an idea of the expected value for x which gives you a place to start looking for the critical value.
$$N \times p = 20 \times 0.3 = 6$$
We need to look at values for x about midway between 6 and 0.
For $x = 3$, $P(X \leq 3) = 0.1070$ which is greater than 0.05
For $x = 2$, $P(X \leq 2) = 0.0355$ which is less than 0.05
Hence the critical value is 2 and the critical region is $X \leq 2$.

$x = 2$ is the first possible value that lies in the critical region. Remember that only integer values of x are allowed.

6 (a) (i) Using $N = 50$, $p = 0.75$ and $x = 31$ with the binomial CD function on the calculator, $P(X \leq 31) = 0.0287$.
$$P(X \geq 44) = 1 - P(X \leq 43)$$
Using $N = 50$, $p = 0.75$ and $x = 43$ with the binomial CD function on the calculator,
$$P(X \leq 43) = 0.9806.$$

$$P(X \geq 44) = 1 - P(X \leq 43) = 1 - 0.9806 = 0.0194$$
$$P(x \leq 31) \cup (x \geq 44) = 0.0287 + 0.0194$$
$$= 0.0481$$
Hence, significance level = 0.0481

(ii) Acceptance region is $32 \leq x \leq 43$

Using $N = 50$, $p = 0.5$ and $x = 43$ using the binomial CD function
$$P(X \leq 43) = 0.9999$$
Using $N = 50$, $p = 0.5$ and $x = 32$ using the binomial CD function
$$P(X \leq 32) = 0.9835$$
$$P(32 \leq x \leq 43) = 0.9999 - 0.9835$$
$$= 0.0164$$

(b) (i) Using $N = 200$, $p = 0.75$ and $x = 139$ using the binomial CD function
$$P(X \leq 139) = 0.0454$$
As this is a two-tailed test we need to multiply the probability by 2 to obtain the p-value.
$$p\text{-value} = 0.0454 \times 2 = 0.0908$$

(ii) p-value $0.0908 > 0.05$ hence we fail to reject H_0 and conclude that there is insufficient evidence to conclude that the probability is not 0.75.

7 (a) (i) If p is the probability of Gwilym winning a game:

Null hypothesis is $\qquad$ $H_0 : p = 0.6$
Alternative hypothesis is $\qquad$ $H_1 : p < 0.6$

(ii) Assuming under the null hypothesis H_0
$$P(X \leq 7) = 0.0210$$
As $0.0210 < 0.05$ there is strong evidence to reject Gwilym's claim.

(b) Under H_0, X is now $B(80, 0.6)$
$$p\text{-value} = P(X \leq 37)$$
$$= 0.008829$$
Now $0.008829 < 0.01$ so there is very strong evidence to reject Gwilym's claim.

8 (a) If p is the probability of a cure using the drug:

Null hypothesis is $\qquad$ $H_0 : p = 0.7$
Alternative hypothesis is $\qquad$ $H_1 : p > 0.7$

(b) If X is the number cured by the drug, X is $B(50, 0.7)$
$$P(X \geq 40) = 1 - P(X \leq 39)$$
$$= 1 - 0.9211$$
$$= 0.0789$$
Now $0.0789 > 0.05$ so there is insufficient evidence to conclude that the new drug is better.

Failing to reject H_0 means that x must lie in range $32 \leq x \leq 43$ so we need to determine the probability that x lies in this range.

Here we have used a calculator to work out $P(X \leq 7)$ using $N = 20$, $p = 0.6$ and $x = 7$ using the binomial CD function.

We therefore fail to reject the null hypothesis H_0. Remember to state the result in the context of the question.

Exam practice answers

If the new drug has not increased the patients cured we need to find the probability of less than or equal to 189 patients cured with the new percentage.

(c) (i) X, the number cured by the drug is now B(250, 0.7)
$$P(X \geq 190) = 1 - P(X \leq 189)$$
$$= 1 - 0.9790$$
$$= 0.021$$
Significance level = 0.021

(ii) X, the number cured by the drug is now B(250, 0.8)
$$P(X \leq 189) = 0.0510$$

Note that this is a two-tailed test.

9 (a) If p is the probability of spinning a six:

Null hypothesis is $\qquad$ $H_0 : p = \frac{1}{6}$

Alternative hypothesis is $\qquad$ $H_1 : p \neq \frac{1}{6}$

(b) If X is the number of sixes the spinner lands on:
$$X \sim B(n, p)$$
$$X \sim B\left(30, \frac{1}{6}\right)$$
$$P(X \geq 9) = 1 - P(X \leq 8)$$
$$= 1 - 0.9494$$
$$= 0.0506$$

Remember to halve the significance level as we are performing a two-tailed test.

Now as $0.0506 > 0.05$ we do not reject the null hypothesis. This means that there is no evidence to suggest the spinner is biased.

(c) Looking at the lower tail first, we need to find the largest value of x such that $P(X \leq x) = 0.05$.

We use the calculator with the binomial CD function to try different values of x with $N = 30$ and $p = \frac{1}{6}$.

$P(X \leq 0) = 0.0042$ (this is less than 0.05 but need to try a higher value to see if it is nearer 0.05).
$P(X \leq 1) = 0.0295$ (this is less than 0.05 and is the highest value of x).
$P(X \leq 2) = 0.1028$ (this is higher than 0.05).

The critical value in the lower tail is $X = 1$ and the critical region is $X \leq 1$.

Looking at the upper tail, we need to find the largest value of x such that $P(X \geq x) = 0.05$.

We use the calculator with the binomial CD function to try different values of x with $N = 30$ and $p = \frac{1}{6}$.

Remember we are looking to find the largest value of x with a probability of ≤ 0.05. In this case the value of x is 10 with a probability of 0.0197.

$$P(X \geq 8) = 1 - P(X \leq 7) = 1 - 0.8863 = 0.1137$$
$$P(X \geq 9) = 1 - P(X \leq 8) = 1 - 0.9494 = 0.0506$$
$$P(X \geq 10) = 1 - P(X \leq 9) = 1 - 0.9803 = 0.0197$$

Hence, the critical regions are $X \leq 1$ and $X \geq 10$.

If a test statistic X were to fall in either of these regions it would lead to the null hypothesis being rejected in favour of the alternative hypothesis.

(d) Probability of a Type I error = The probability of rejecting the null hypothesis when it is true.
$$= P(X \le 1) \text{ or } P(X \ge 10)$$
$$= 0.0295 + 0.0197$$
$$= 0.0492$$

(e) Notice that the probability, p, is now 0.5.
Probability of a Type II error = The probability of not rejecting the null hypothesis when it is false.
$$= P(2 \le X \le 9)$$
$$= P(X \le 9) - P(X \le 1)$$
$$= 0.0214 - 0.0000$$
$$= 0.0214$$

Note that these are the two probabilities of rejecting the null hypothesis.

Watch out

Make sure you don't use $P(X \le 2)$ here.

Topic 15

1 (a) Using $s = \frac{1}{2}(u+v)t$
$$720 = \frac{1}{2}(20+25)t$$
$$t = \frac{1440}{45}$$
$$= 32 \text{ s}$$

As the acceleration is constant, the equations of motion can be used.

(b)

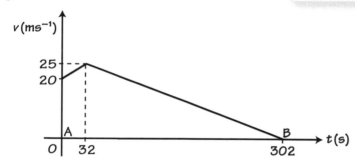

Always look back at what you have already found and add it to your graph. Here we know that the time taken during acceleration is 32 s.

(c) Distance between A and B = area under the velocity–time graph.
$$= \text{area of trapezium} + \text{area of triangle}$$
$$= \frac{1}{2}(20+25)32 + \frac{1}{2} \times 270 \times 25$$
$$= 4095 \text{ m}$$

2 (a) $v^2 = u^2 + 2as$
$$= (0)^2 + 2 \times 9.8 \times 140$$
$$v = 52.38 \text{ m s}^{-1}$$

(b) $v = u + at$
$$t = \frac{v-u}{a} = \frac{52.38 - 0}{9.8} = 5.34 \text{ s}$$

(c) The object is modelled as a particle or air resistance is ignored.

The acceleration will be constant at g provided that air resistance can be neglected.

Exam practice answers

Notice in the equation $v = 64 - \frac{1}{27}t^3$, v depends on t.

This means the velocity and hence the acceleration varies so you cannot use the equations of motion. Instead we need to use calculus.

Always look for some information in the question that would allow you to find the value for the constant of integration.

3 (a) $v = 64 - \frac{1}{27}t^3$

When the car comes to rest, $v = 0$
so we have:

$$0 = 64 - \frac{1}{27}t^3$$

Hence $\frac{1}{27}t^3 = 64$

$$t^3 = 64 \times 27$$

Cube-rooting both sides gives $t = 12$ s

(b) $r = \int v\,dt$

$$r = \int \left(64 - \frac{1}{27}t^3\right)dt$$

$$= 64t - \frac{1}{108}t^4 + c$$

When $t = 0$, $r = 0$
$$0 = 0 - 0 + c$$
$$c = 0$$

when $t = 12$, $r = 64 \times 12 - \frac{1}{108} \times 12^4$

Distance PQ = 576 m

4 (a) $2t^2 - 7t + 5 = 0$
$$(t - 1)(2t - 5) = 0$$
$$t = 1\text{ s or } 2.5\text{ s}$$

(b) $r = \int v\,dt$

$$r = \int \left(2t^2 - 7t + 5\right)dt$$

$$= \frac{2t^3}{3} - \frac{7t^2}{2} + 5t + c$$

$r = 0$ when $t = 1$

$$0 = \frac{2(1)^3}{3} - \frac{7(1)^2}{2} + 5(1) + c$$

$$0 = \frac{2}{3} - \frac{7}{2} + 5 + c$$

$$c = -\frac{13}{6}$$

The value of the constant c is now substituted back into the equation.

Hence, $r = \frac{2t^3}{3} - \frac{7t^2}{2} + 5t - \frac{13}{6}$

(c) $a = \frac{dv}{dt}$

$$= 4t - 7$$

5 (a) $a = -9.8\,\mathrm{m\,s^{-2}}$, $u = 14.7\,\mathrm{m\,s^{-1}}$, $t = 2\,\mathrm{s}$

$$v = u + at$$
$$= 14.7 + (-9.8)(2)$$
$$= -4.9\,\mathrm{m\,s^{-1}}\ \text{(this is a downwards velocity)}$$

Hence, speed $= 4.9\,\mathrm{m\,s^{-1}}$

(b) Taking the upward direction as positive:

$a = -9.8\,\mathrm{m\,s^{-2}}$, $u = 14.7\,\mathrm{m\,s^{-1}}$, $s = -70.2\,\mathrm{m}$

$$v^2 = u^2 + 2as$$
$$= (14.7)^2 + 2(-9.8)(-70.2) = 1592.01$$
$$v = 39.9\,\mathrm{m\,s^{-1}}$$

(c) $s = 3.969\,\mathrm{m}$, $a = -9.8\,\mathrm{m\,s^{-2}}$, $u = 14.7\,\mathrm{m\,s^{-1}}$

$$s = ut + \tfrac{1}{2}at^2$$

$$3.969 = 14.7t + \tfrac{1}{2}\left(-9.8\right)t^2$$

$t^2 - 3t + 0.81 = 0$
$(t - 2.7)(t - 0.3) = 0$
$t = 2.7$ or $0.3\,\mathrm{s}$
Required length of time $= 2.7 - 0.3 = 2.4\,\mathrm{s}$

> Gravity opposes the upward motion, so we need a negative value for g.

> We have taken the upward direction as positive, so a negative sign indicates the velocity (v in this case) is in the downward direction.

6 (a) (i) Let x = the height from the ground where they collide.
For the ball dropped from the tower:
$u = 0$, $a = g = 9.8$, $s = 24 - x$

$$s = ut + \tfrac{1}{2}at^2$$

$$24 - x = 0 + \tfrac{1}{2}\left(9.8\right)t^2$$

$$24 - x = 4.9t^2$$
$$x = 24 - 4.9t^2 \qquad (1)$$

For the ball thrown upwards from the ground we have
$u = 15$, $g = -9.8$, $s = x$

$$s = ut + \tfrac{1}{2}at^2$$

$$x = 15t + \tfrac{1}{2}\left(-9.8\right)t^2$$

$$= 15t - 4.9t^2 \qquad (2)$$

Equating equations (1) and (2)
$$24 - 4.9t^2 = 15t - 4.9t^2$$
$$t = 1.6\,\mathrm{s}$$

(ii) When $t = 1.6\,\mathrm{s}$,
$$x = 24 - 4.9t^2 = 24 - 4.9(1.6)^2 = 11.456\,\mathrm{m}$$

(b) The balls are modelled as particles or no air resistance/friction.

> Note that when they collide, they will be the same distance from the ground.

> Since the distances above the ground are the same at the point of collision, we can equate equations (1) and (2) to find the time, t.

7 (a) Velocity between O and A = gradient $= \dfrac{10}{5} = 2\,\mathrm{m\,s^{-1}}$

(b) Velocity between B and D = gradient $= -\dfrac{20}{4} = -5\,\mathrm{m\,s^{-1}}$

(c) (i) Total distance $= 10 + 10 + 10 = 30\,\mathrm{m}$
(ii) Displacement $= 10 - 10 - 10 = -10\,\mathrm{m}$

Topic 16

1 (a)

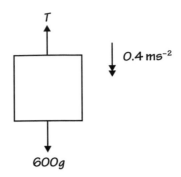

Applying Newton's 2nd law in the vertical direction, we have:
$$600 \times 0.4 = 600g - T$$
$$240 = 600 \times 9.8 - T$$
$$T = 5640 \text{ N}$$

(b) When the lift is travelling at constant speed, there is no resultant force to produce an acceleration. In this situation, the tension in the cable will equal the weight.
$$T = mg$$
$$T = 600 \times 9.8$$
$$= 5880 \text{ N}$$

2 (a)

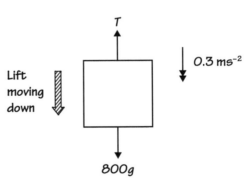

Applying Newton's 2nd law in the vertical direction, we have:
$$800 \times 0.3 = 800g - T$$
$$240 = 800 \times 9.8 - T$$
$$T = 7600 \text{ N}$$

(b)

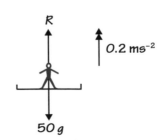

Applying Newton's 2nd law in the vertical direction for the man, we have:
$$50 \times 0.2 = R - 50g$$
$$10 = R - 50 \times 9.8$$
$$R = 500 \text{ N}$$

(c) The lift/man is modelled as a particle.

3

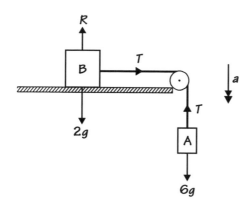

Applying Newton's 2nd law to mass B:
$$2a = T$$
Applying Newton's 2nd law to mass A:
$$6a = 6g - T$$
$$6a = 6 \times 9.8 - T$$
$$6a = 58.8 - T$$
Now $T = 2a$ so $6a = 58.8 - 2a$
$$8a = 58.8$$
$$a = 7.35 \text{ m s}^{-2}$$
$$T = 2a = 14.7 \text{ N}$$

4 (a)

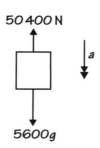

Applying Newton's 2nd law in the vertical direction, we have:
$$5600 \times a = 5600g - 50\,400$$
$$5600 \times a = 5600 \times 9.8 - 50\,400$$
$$a = 0.8 \text{ m s}^{-2}$$

(b) $v = u + at$
$$= 0 + 0.8 \times 8$$
$$= 6.4 \text{ m s}^{-1}$$

(c)

Velocity
(ms^{-1})

6.4

0 8   33   40   Time (s)

(d) Distance travelled = area under the velocity–time graph

$$= \tfrac{1}{2}\left(25 + 40\right)6.4$$
$$= 208\,\text{m}$$

(e) The maximum tension occurs when it is decelerating to rest.

$$\text{Deceleration} = \frac{v - u}{t} = \frac{0 - 6.4}{(40 - 33)} = -0.9143\,\text{m s}^{-2}$$

Applying Newton's 2nd law in the vertical direction, we have:

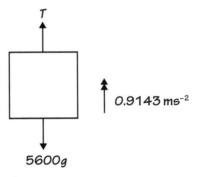

$$5600 \times 0.9143 = T - 5600 \times 9.8$$
$$T = 60\,000\,\text{N}$$

> Here we will take the upward direction as positive.

Topic 17

1 Resultant force = **P** + **Q** + **R**
$$= 3\mathbf{i} + \mathbf{j} + 11\mathbf{i} - 4\mathbf{j} - 2\mathbf{i} + 8\mathbf{j}$$
$$= 12\mathbf{i} + 5\mathbf{j}$$

Magnitude of the resultant force = $\sqrt{12^2 + 5^2}$
$$= 13\,\text{N}$$

Magnitude of acceleration $= \dfrac{F}{m} = \dfrac{13}{5} = 2.6\,\text{N}$

2 (a) Resultant force = $\mathbf{F}_1 + \mathbf{F}_2$
$$= (3\mathbf{i} + 5\mathbf{j}) + (5\mathbf{i} - \mathbf{j})$$
$$= 8\mathbf{i} + 4\mathbf{j}$$

$$\theta = \tan^{-1}\left(\frac{4}{8}\right) = 26.6°$$

(b) Magnitude of the force $= \sqrt{8^2 + 4^2}$

$$= \sqrt{80}$$

$$= 4\sqrt{5}$$

3 (a) $\mathbf{F} + \mathbf{G} + \mathbf{H} = 0$

$$4\mathbf{i} - 2\mathbf{j} + \mathbf{i} + 7\mathbf{j} + a\mathbf{i} - b\mathbf{j} = 0$$

$$(5 + a)\mathbf{i} + (5 - b)\mathbf{j} = 0$$

Hence $a = -5$ and $b = 5$

$$\mathbf{H} = -5\mathbf{i} - 5\mathbf{j}$$

Magnitude of $\mathbf{H} = \sqrt{(-5)^2 + (-5)^2}$

$$= \sqrt{50}$$

$$= 5\sqrt{2}\ \text{N}$$

(b) $a = \dfrac{F}{m}$

$$= \frac{5\sqrt{2}}{5}$$

$$= \sqrt{2}\ \text{m s}^{-2}$$

4 (a)

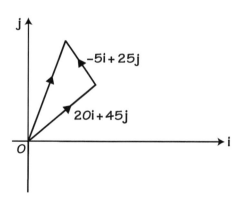

Total displacement $= 20\mathbf{i} + 45\mathbf{j} - 5\mathbf{i} + 25\mathbf{j} = 15\mathbf{i} + 70\mathbf{j}$

(b) Total distance travelled $= \sqrt{20^2 + 45^2} + \sqrt{(-5)^2 + 25^2}$

$$= \sqrt{2425} + \sqrt{650}$$

$$= 74.7\ \text{km (3 s.f.)}$$

5 (a) $\mathbf{F} + \mathbf{R} = 5\mathbf{i} + 2\mathbf{j}$

$$2\mathbf{i} - 3\mathbf{j} + a\mathbf{i} + b\mathbf{j} = 5\mathbf{i} + 2\mathbf{j}$$

Hence $a = 3$ and $b = 5$

(b) Resultant before $\mathbf{Q}$ added $= 5\mathbf{i} + 2\mathbf{j}$

As resultant becomes zero, $\mathbf{Q} = -5\mathbf{i} - 2\mathbf{j}$

6 (a) Resultant force $= \begin{pmatrix} -1 \\ 4 \end{pmatrix} + \begin{pmatrix} 3 \\ 2 \end{pmatrix} = \begin{pmatrix} 2 \\ 6 \end{pmatrix}$

Magnitude of resultant force $= \sqrt{2^2 + 6^2} = \sqrt{40} = 2\sqrt{10}$

Acceleration $= \dfrac{F}{m} = \dfrac{2\sqrt{10}}{10} = \dfrac{\sqrt{10}}{5} \, \mathrm{m\,s^{-2}}$

(b) $s = ut + \frac{1}{2}at^2$

$= 0 + \dfrac{1}{2} \times \left(\dfrac{\sqrt{10}}{5} \right) \times 3^2$

$= 2.85 \, \mathrm{m}$ (2 d.p.)